Basic Techniques in

BIOCHEMISTRY AND MOLECULAR BIOLOGY

Basic Techniques in

BIOCHEMISTRY AND MOLECULAR BIOLOGY

Dr. Sarfaraz Ahmad

ANMOL PUBLICATIONS PVT. LTD.
NEW DELHI-110 002 (INDIA)

ANMOL PUBLICATIONS PVT. LTD.
Regd. Office: 4360/4, Ansari Road, Daryaganj,
New Delhi-110002 (India)
Tel.: 23278000, 23261597, 23286875, 23255577
Fax: 91-11-23280289
Email: anmolpub@gmail.com
Visit us at: www.anmolpublications.com

Branch Office: No. 1015, Ist Main Road, BSK IIIrd Stage
IIIrd Phase, IIIrd Block, Bengaluru-560 085 (India)
Tel.: 080-41723429 · Fax: 080-26723604
Email: anmolpublicationsbangalore@gmail.com

Basic Techniques in Biochemistry and Molecular Biology

First Edition, 2011

ISBN 978-81-261-4730-4

PRINTED IN INDIA

Printed at Mehra Offset Press, Delhi.

Contents

Preface

Most of the Universe consists of matter and energy. Energy is the capacity to do work. Matter has mass and occupies space. All matter is composed of basic elements that cannot be broken down to substances with different chemical or physical properties. Elements are substances consisting of one type of atom, for example Carbon atoms make up diamond, and also graphite. Pure (24K) gold is composed of only one type of atom, gold atoms. Atoms are the smallest particle into which an element can be divided. The ancient Greek philosophers developed the concept of the atom, although they considered it the fundamental particle that could not be broken down. Since the work of Enrico Fermi and his colleagues, we now know that the atom is divisible, often releasing tremendous energies as in nuclear explosions or (in a controlled fashion in) thermonuclear power plants.

Biochemistry, sometimes abbreviated as "BioChem", is the study of chemical processes in living organisms. Biochemistry governs all living organisms and living processes. By controlling information flow through biochemical signalling and the flow of chemical energy through metabolism; biochemical processes give rise to the seemingly magical phenomenon of life. Much of biochemistry deals with the structures and functions of cellular components such as proteins, carbohydrates, lipids, nucleic acids and other biomolecules although increasingly processes rather than individual molecules are the main focus. Over the last 40 years biochemistry has become so successful at explaining living processes that now almost all areas of the life sciences from botany to medicine are engaged in biochemical research. Today the main focus of pure biochemistry is in understanding how biological molecules give rise to the processes that occur within living cells which in turn relates greatly to the study and understanding of whole organisms.

The dawn of biochemistry may have been the discovery of the first enzyme, diastase (today called amylase), in 1833 by Anselme Payen. Eduard Buchner contributed the first demonstration of a complex biochemical process outside of a cell in 1896: alcoholic fermentation in cell extracts of yeast. Although the term "biochemistry" seems to have been first used in

1882, it is generally accepted that the formal coinage of biochemistry occurred in 1903 by Carl Neuberg, a German chemist. Previously, this area would have been referred to as physiological chemistry. Since then, biochemistry has advanced, especially since the mid-20th century, with the development of new techniques such as chromatography, X-ray diffraction, dual polarisation interferometry, NMR spectroscopy, radioisotopic labelling, electron microscopy and molecular dynamics simulations. These techniques allowed for the discovery and detailed analysis of many molecules and metabolic pathways of the cell, such as glycolysis and the Krebs cycle (citric acid cycle).

Another significant historic event in biochemistry is the discovery of the gene and its role in the transfer of information in the cell. This part of biochemistry is often called molecular biology. In the 1950s, James D. Watson, Francis Crick, Rosalind Franklin, and Maurice Wilkins were instrumental in solving DNA structure and suggesting its relationship with genetic transfer of information. In 1988, Colin Pitchfork was the first person convicted of murder with DNA evidence, which led to growth of forensic science. More recently, Andrew Z. Fire and Craig C. Mello received the 2006 Nobel Prize for discovering the role of RNA interference (RNAi), in the silencing of gene expression. Today, there are three main types of biochemistry. Plant biochemistry involves the study of the biochemistry of autotrophic organisms such as photosynthesis and other plant specific biochemical processes. General biochemistry encompasses both plant and animal biochemistry. Human/medical/medicinal biochemistry focuses on the biochemistry of humans and medical illnesses.

The aim of this book is to present a core of biochemical knowledge that is desirable for undergraduate and postgraduate students and also those involved in the field of medical, microbiology, biotechnology and pharmaceutical. Every attempt has been made to keep abrest of the advances in the subject and at the same time to include the fundamentals.

—Author

1

Introduction

Biochemistry, sometimes abbreviated as "BioChem", is the study of chemical processes in living organisms. Biochemistry governs all living organisms and living processes. By controlling information flow through biochemical signalling and the flow of chemical energy through metabolism; biochemical processes give rise to the seemingly magical phenomenon of life. Much of biochemistry deals with the structures and functions of cellular components such as proteins, carbohydrates, lipids, nucleic acids and other biomolecules although increasingly processes rather than individual molecules are the main focus. Over the last 40 years biochemistry has become so successful at explaining living processes that now almost all areas of the life sciences from botany to medicine are engaged in biochemical research. Today the main focus of pure biochemistry is in understanding how biological molecules give rise to the processes that occur within living cells which in turn relates greatly to the study and understanding of whole organisms.

Among the vast number of different biomolecules, many are complex and large molecules (called *polymers*), which are composed of similar repeating subunits (called *monomers*). Each class of polymeric biomolecule has a different set of subunit types. For example, a protein is a polymer whose subunits are selected from a set of 20 or more amino acids. Biochemistry studies the chemical properties of important biological molecules, like proteins, and in particular the chemistry of enzyme-catalysed reactions. The biochemistry of cell metabolism and the endocrine system has been extensively described. Other areas of biochemistry include the genetic code (DNA, RNA), protein synthesis, cell membrane transport, and signal transduction.

History

Originally, it was generally believed that life was not subject to the laws of science the way non-life was. It was thought that only living beings could produce the molecules of life (from other, previously existing biomolecules). Then, in 1828, Friedrich Wöhler published a paper on the synthesis of urea, proving that organic compounds can be created artificially.

The dawn of biochemistry may have been the discovery of the first enzyme, diastase (today called amylase), in 1833 by Anselme Payen. Eduard Buchner contributed the first demonstration of a complex biochemical process outside of a cell in 1896: alcoholic fermentation in cell extracts of yeast. Although the term "biochemistry" seems to have been first used in 1882, it is generally accepted that the formal coinage of biochemistry occurred in 1903 by Carl Neuberg, a German chemist.

Previously, this area would have been referred to as physiological chemistry. Since then, biochemistry has advanced, especially since the mid-20th century, with the development of new techniques such as chromatography, X-ray diffraction, dual polarisation interferometry, NMR spectroscopy, radioisotopic labelling, electron microscopy and molecular dynamics simulations. These techniques allowed for the discovery and detailed analysis of many molecules and metabolic pathways of the cell, such as glycolysis and the Krebs cycle (citric acid cycle).

Another significant historic event in biochemistry is the discovery of the gene and its role in the transfer of information in the cell. This part of biochemistry is often called molecular biology. In the 1950s, James D. Watson, Francis Crick, Rosalind Franklin, and Maurice Wilkins were instrumental in solving DNA structure and suggesting its relationship with genetic transfer of information. In 1988, Colin Pitchfork was the first person convicted of murder with DNA evidence, which led to growth of forensic science. More recently, Andrew Z. Fire and Craig C. Mello received the 2006 Nobel Prize for discovering the role of RNA interference (RNAi), in the silencing of gene expression. Today, there are three main types of biochemistry. Plant biochemistry involves the study of the biochemistry of autotrophic organisms such as photosynthesis and other plant specific biochemical processes. General biochemistry encompasses both plant and animal biochemistry. Human/medical/medicinal biochemistry focuses on the biochemistry of humans and medical illnesses.

Monomers and Polymers

The four main classes of molecules in biochemistry are carbohydrates, lipids, proteins, and nucleic acids. Many biological molecules are polymers: in this terminology, *monomers* are relatively small micromolecules that are linked together to create large macromolecules, which are known as *polymers*. When monomers are linked together to synthesize a biological polymer, they undergo a process called dehydration synthesis.

Carbohydrates

Carbohydrates are made from monomers called *monosaccharides.* Some of these monosaccharides include glucose ($C_6H_{12}O_6$), fructose ($C_6H_{12}O_6$), and deoxyribose ($C_5H_{10}O_4$). When two monosaccharides undergo dehydration synthesis, water is produced, as two hydrogen atoms and one oxygen atom are lost from the two monosaccharides' hydroxyl group.

Lipids

Lipids are usually made from one molecule of glycerol combined with other molecules. In triglycerides, the main group of bulk lipids, there is one molecule of glycerol and three fatty acids. Fatty acids are considered the monomer in that case, and may be saturated (no double bonds in the carbon chain) or unsaturated (one or more double bonds in the carbon chain). Lipids, especially phospholipids, are also used in various pharmaceutical products, either as co-solubilisers (e.g. in parenteral infusions) or else as drug carrier components (e.g. in a liposome or transfersome).

Proteins

Proteins are very large molecules – macro-biopolymers – made from monomers called *amino acids.* There are 20 standard amino acids, each containing a carboxyl group, an amino group, and a side chain (known as an "R" group). The "R" group is what makes each amino acid different, and the properties of the side chains greatly influence the overall three-dimensional conformation of a protein. When amino acids combine, they form a special bond called a peptide bond through dehydration synthesis, and become a *polypeptide,* or protein.

Nucleic Acids

Nucleic acids are the molecules that make up DNA, an extremely important substance which all cellular organisms use to store their genetic information. The most common nucleic acids are deoxyribonucleic acid and ribonucleic acid. Their monomers are called nucleotides. The most common nucleotides are adenine, cytosine, guanine, thymine, and uracil. Adenine binds with thymine and uracil; thymine only binds with adenine; and cytosine and guanine can only bind with each other.

Carbohydrates

The function of carbohydrates includes energy storage and providing structure. Sugars are carbohydrates, but not all carbohydrates are sugars. There are more carbohydrates on Earth than any other known type of biomolecule; they are used to store energy and genetic information, as well as play important roles in cell to cell interactions and communications.

Monosaccharides

The simplest type of carbohydrate is a monosaccharide, which among other properties contains carbon, hydrogen, and oxygen, mostly in a ratio of 1:2:1 (generalized formula $C_nH_{2n}O_n$, where n is at least 3). Glucose, one of the most important carbohydrates, is an example of a monosaccharide. So is fructose, the sugar commonly associated with the sweet taste of fruits. Some carbohydrates (especially after condensation to oligo-and polysaccharides) contain less carbon relative to H and O, which still are present in 2:1 (H:O) ratio. Monosaccharides can be grouped into aldoses (having an aldehyde group at the end of the chain, e.g. glucose) and ketoses (having a keto group in their chain; e.g. fructose). Both aldoses and ketoses occur in an equilibrium (starting with chain lengths of C4) cyclic forms. These are generated by bond formation between one of the hydroxyl groups of the sugar chain with the carbon of the aldehyde or keto group to form a hemiacetal bond. This leads to saturated five-membered (in furanoses) or six-membered (in pyranoses) heterocyclic rings containing one O as heteroatom.

Disaccharides

Two monosaccharides can be joined together using dehydration synthesis, in which a hydrogen atom is removed from the end of one molecule and a hydroxyl group (—OH) is removed from the other; the remaining residues are then attached at the sites from which the atoms were removed. The H—OH or H_2O is then released as a molecule of water, hence the term *dehydration*. The new molecule, consisting of two monosaccharides, is called a *disaccharide* and is conjoined together by a glycosidic or ether bond.

The reverse reaction can also occur, using a molecule of water to split up a disaccharide and break the glycosidic bond; this is termed *hydrolysis*. The most well-known disaccharide is sucrose, ordinary sugar (in scientific contexts, called *table sugar* or *cane sugar* to differentiate it from other sugars). Sucrose consists of a glucose molecule and a fructose molecule joined together. Another important disaccharide is lactose, consisting of a glucose molecule and a galactose molecule. As most humans age, the production of lactase, the enzyme that hydrolyzes lactose back into glucose and galactose, typically decreases. This results in lactase deficiency, also called *lactose intolerance*.

Sugar polymers are characterised by having reducing or non-reducing ends. A reducing end of a carbohydrate is a carbon atom which can be in equilibrium with the open-chain aldehyde or keto form. If the joining of monomers takes place at such a carbon atom, the free hydroxy group of the pyranose or furanose form is exchanged with an OH-side chain of another sugar, yielding a full acetal.

This prevents opening of the chain to the aldehyde or keto form and renders the modified residue non-reducing. Lactose contains a reducing end at its glucose moiety, whereas the galactose moiety form a full acetal with the C4-OH group of glucose. Saccharose does not have a reducing end because of full acetal formation between the aldehyde carbon of glucose (C1) and the keto carbon of fructose (C2).

Oligosaccharides and Polysaccharides

When a few (around three to six) monosaccharides are joined together, it is called an *oligosaccharide* (*oligo-* meaning "few"). These molecules tend to be used as markers and signals, as well as having some other uses. Many monosaccharides joined together make a polysaccharide. They can be joined together in one long linear chain, or they may be branched. Two of the most common polysaccharides are cellulose and glycogen, both consisting of repeating glucose monomers.

- *Cellulose* is made by plants and is an important structural component of their cell walls. Humans can neither manufacture nor digest it.
- *Glycogen,* on the other hand, is an animal carbohydrate; humans and other animals use it as a form of energy storage.

Use of Carbohydrates as an Energy Source

Glucose is the major energy source in most life forms. For instance, polysaccharides are broken down into their monomers (glycogen phosphorylase removes glucose residues from glycogen). Disaccharides like lactose or sucrose are cleaved into their two component monosaccharides.

Glycolysis (Anaerobic)

Glucose is mainly metabolized by a very important ten-step pathway called glycolysis, the net result of which is to break down one molecule of glucose into two molecules of pyruvate; this also produces a net two molecules of ATP, the energy currency of cells, along with two reducing equivalents in the form of converting NAD^+ to NADH. This does not require oxygen; if no oxygen is available (or the cell cannot use oxygen), the NAD is restored by converting the pyruvate to lactate (lactic acid) (e.g. in humans) or to ethanol plus carbon dioxide (e.g. in yeast). Other monosaccharides like galactose and fructose can be converted into intermediates of the glycolytic pathway.

Aerobic

In aerobic cells with sufficient oxygen, like most human cells, the pyruvate is further metabolized. It is irreversibly converted to acetyl-CoA, giving off one carbon atom as the waste product carbon dioxide, generating another reducing equivalent as NADH. The two molecules acetyl-CoA (from one molecule of glucose) then enter the citric acid cycle, producing two more molecules of ATP, six more NADH molecules and two reduced (ubi)quinones (via $FADH_2$ as enzyme-bound cofactor), and releasing the remaining carbon atoms as carbon dioxide. The produced NADH and quinol molecules then feed into the enzyme complexes of the respiratory chain, an electron transport system transferring the electrons ultimately to oxygen and conserving the released energy in the form of a proton gradient over a membrane (inner mitochondrial membrane in eukaryotes).

Thereby, oxygen is reduced to water and the original electron acceptors NAD^+ and quinone are regenerated. This is why humans breathe in oxygen and breathe out carbon dioxide. The energy released from transferring the electrons from high-energy states in NADH and quinol is conserved first as proton gradient and converted to ATP via ATP synthase. This generates an additional *28* molecules of ATP (24 from the 8 NADH + 4 from the 2 quinols), totalling to 32 molecules of ATP conserved per degraded glucose (two from glycolysis + two from the citrate cycle). It is clear that using oxygen to completely oxidize glucose provides an organism with far more energy than any oxygen-independent metabolic feature, and this is thought to be the reason why complex life appeared only after Earth's atmosphere accumulated large amounts of oxygen.

Gluconeogenesis

In vertebrates, vigorously contracting skeletal muscles (during weightlifting or sprinting, for example) do not receive enough oxygen to meet the energy demand, and so they shift to anaerobic metabolism, converting glucose to lactate. The liver regenerates the glucose, using a process called gluconeogenesis. This process is not quite the opposite of glycolysis, and actually requires three times the amount of energy gained from glycolysis (six molecules of ATP are used, compared to the two gained in glycolysis). Analogous to the above reactions, the glucose produced can then undergo glycolysis in tissues that need energy, be stored as glycogen (or starch in plants), or be converted to other monosaccharides or joined into di- or oligosaccharides. The combined pathways of glycolysis during exercise, lactate's crossing via the bloodstream to the liver, subsequent gluconeogenesis and release of glucose into the bloodstream is called the Cori cycle.

Proteins

Like carbohydrates, some proteins perform largely structural roles. For instance, movements of the proteins actin and myosin ultimately are responsible for the contraction of skeletal muscle. One property many proteins have is that they specifically bind to a certain molecule or class of molecules—they may be *extremely* selective in what they bind. Antibodies are an example of proteins that attach to one specific type of molecule. In fact, the enzyme-linked immunosorbent assay (ELISA), which uses antibodies, is currently one of the most sensitive tests modern medicine uses to detect various biomolecules. Probably the most important proteins, however, are the enzymes. These molecules recognize specific reactant molecules called *substrates*; they then catalyse the reaction between them. By lowering the activation energy, the enzyme speeds up that reaction by a rate of 10^{11} or more: a reaction that would normally take over 3,000 years to complete spontaneously might take less than a second with an enzyme. The enzyme itself is not used up in the process, and is free to catalyse the same reaction with a new set of substrates. Using various modifiers, the activity of the enzyme can be regulated, enabling control of the biochemistry of the cell as a whole.

In essence, proteins are chains of amino acids. An amino acid consists of a carbon atom bound to four groups. One is an amino group, $—NH_2$, and one is a carboxylic acid group, —COOH (although these exist as $—NH_3^+$ and $—COO^-$ under physiologic conditions). The third is a simple hydrogen atom. The fourth is commonly denoted "—R" and is different for each amino acid. There are twenty standard amino acids. Some of these have functions by themselves or in a modified form; for instance, glutamate functions as an important neurotransmitter.

Amino acids can be joined together via a peptide bond. In this dehydration synthesis, a water molecule is removed and the peptide bond connects the nitrogen of one amino acid's amino group to the carbon of the other's carboxylic acid group. The resulting molecule is called a *dipeptide*, and short stretches of amino acids (usually, fewer than around thirty) are called *peptides* or polypeptides. Longer stretches merit the title *proteins*. As an example, the important blood serum protein albumin contains 585 amino acid residues.

The structure of proteins is traditionally described in a hierarchy of four levels. The primary structure of a protein simply consists of its linear sequence of amino acids; for instance, "alanine-glycine-tryptophan-serine-glutamate-asparagine-glycine-lysine-...". Secondary structure is concerned with local morphology (morphology being the study of structure). Some combinations of amino acids will tend to curl up in a coil called an α-helix or into a sheet called a β-sheet; some α-helixes can be seen in the hemoglobin schematic above. Tertiary structure is the entire three-dimensional shape of the protein. This shape is determined by the sequence of amino acids. In fact, a single change can change the entire structure. The alpha chain of hemoglobin contains 146 amino acid residues; substitution of the glutamate residue at position 6 with a valine residue changes the behaviour of hemoglobin so much that it results in sickle-cell disease. Finally quaternary structure is concerned with the structure of a protein with multiple peptide subunits, like hemoglobin with its four subunits. Not all proteins have more than one subunit.

Ingested proteins are usually broken up into single amino acids or dipeptides in the small intestine, and then absorbed. They can then be joined together to make new proteins. Intermediate products of glycolysis, the citric acid cycle, and the pentose phosphate pathway can be used to make all twenty amino acids, and most bacteria and plants possess all the necessary enzymes to synthesize them. Humans and other mammals, however, can only synthesize half of them. They cannot synthesize isoleucine, leucine, lysine, methionine, phenylalanine, threonine, tryptophan, and valine. These are the essential amino acids, since it is essential to ingest them. Mammals do possess the enzymes to synthesize alanine, asparagine, aspartate, cysteine, glutamate, glutamine, glycine, proline, serine, and tyrosine, the nonessential amino acids. While they can synthesize arginine and histidine, they cannot produce it in sufficient amounts for young, growing animals, and so these are often considered essential amino acids.

If the amino group is removed from an amino acid, it leaves behind a carbon skeleton called an α-keto acid. Enzymes called transaminases can easily transfer the amino group from one amino acid (making it an α-keto acid) to another α-keto acid (making it an amino acid). This is important in the biosynthesis of amino acids, as for many of the pathways, intermediates from other biochemical pathways are converted to the α-keto acid skeleton, and then an amino group is added, often via transamination. The amino acids may then be linked together to make a protein.

A similar process is used to break down proteins. It is first hydrolysed into its component amino acids. Free ammonia (NH_3), existing as the ammonium ion (NH_4^+) in blood, is toxic to life forms. A suitable method for excreting it must therefore exist. Different strategies have evolved in different animals, depending on the animals' needs. Unicellular organisms, of course, simply release the ammonia into the environment. Similarly, bony fish can release the ammonia into the water where it is quickly diluted. In general, mammals convert the ammonia into urea, via the urea cycle.

Lipids

The term lipid comprises a diverse range of molecules and to some extent is a catchall for relatively water-insoluble or nonpolar compounds of biological origin, including waxes, fatty acids, fatty-acid derived phospholipids, sphingolipids, glycolipids and terpenoids (e.g. retinoids and steroids). Some lipids are linear aliphatic molecules, while others have ring structures. Some are aromatic, while others are not. Some are flexible, while others are rigid.

Most lipids have some polar character in addition to being largely nonpolar. Generally, the bulk of their structure is nonpolar or hydrophobic ("water-fearing"), meaning that it does not interact well with polar solvents like water. Another part of their structure is polar or hydrophilic ("water-loving") and will tend to associate with polar solvents like water. This makes them amphiphilic molecules (having both hydrophobic and hydrophilic portions). In the case of cholesterol, the polar group is a mere -OH (hydroxyl or alcohol). In the case of phospholipids, the polar groups are considerably larger and more polar. Lipids are an integral part of our daily diet. Most oils and milk products that we use for cooking and eating like butter, cheese, ghee etc., are composed of fats. Vegetable oils are rich in various polyunsaturated fatty acids (PUFA). Lipid-containing foods undergo digestion within the body and are broken into fatty acids and glycerol, which are the final degradation products of fats and lipids.

Nucleic Acids

A nucleic acid is a complex, high-molecular-weight biochemical macromolecule composed of nucleotide chains that convey genetic information. The most common nucleic acids are deoxyribonucleic acid (DNA) and ribonucleic acid (RNA). Nucleic acids are found in all living cells and viruses. Aside from the genetic material of the cell, nucleic acids often play a role as second messengers, as well as forming the base molecule for adenosine triphosphate, the primary energy-carrier molecule found in all living organisms.

Nucleic acid, so called because of its prevalence in cellular nuclei, is the generic name of the family of biopolymers. The monomers are called nucleotides, and each consists of three components: a nitrogenous heterocyclic base (either a purine or a pyrimidine), a pentose sugar, and a phosphate group. Different nucleic acid types differ in the specific sugar found in their chain (e.g. DNA or deoxyribonucleic acid contains 2-deoxyriboses). Also, the nitrogenous bases possible in the two nucleic acids are different: adenine, cytosine, and guanine occur in both RNA and DNA, while thymine occurs only in DNA and uracil occurs in RNA.

Relationship to other "Molecular-scale" Biological Sciences

Researchers in biochemistry use specific techniques native to biochemistry, but increasingly combine these with techniques and ideas from genetics, molecular biology and biophysics. There has never been a hard-line between these disciplines in terms of content and technique. Today the terms *molecular biology* and *biochemistry* are nearly interchangeable.

- *Biochemistry* is the study of the chemical substances and vital processes occurring in living organisms. Biochemists focus heavily on the role, function, and structure of biomolecules. The study of the chemistry behind biological processes and the synthesis of biologically active molecules are examples of biochemistry.
- *Genetics* is the study of the effect of genetic differences on organisms. Often this can be inferred by the absence of a normal component (e.g. one gene). The study of "mutants" – organisms which lack one or more functional components with respect to the so-called "wild type" or normal phenotype. Genetic interactions (epistasis) can often confound simple interpretations of such "knock-out" studies.
- *Molecular biology* is the study of molecular underpinnings of the process of replication, transcription and translation of the genetic material. The central dogma of molecular biology where genetic material is transcribed into RNA and then translated into protein, despite being an oversimplified picture of molecular biology, still provides a good starting point for understanding the field. This picture, however, is undergoing revision in light of emerging novel roles for RNA.
- *Chemical Biology* seeks to develop new tools based on small molecules that allow minimal perturbation of biological systems while providing detailed information about their function. Further, chemical biology employs biological systems to create non-natural hybrids between biomolecules and synthetic devices.

Hydrogen Ion Concentration and Buffers

The hydrogen ion consists of a single positively charged particle (the proton) that is not orbited by any electrons. The hydrogen ion is, therefore, the smallest ionic particle and is extremely reactive. It is this fact that accounts for its profound effect on the functioning of biological systems at very low concentrations.

In the environment hydrogen ion concentrations vary over an enormous scale (from less than 10-14 mol/l to more than 1mol/l).

Table 1 gives examples of pH values and corresponding hydrogen ion concentrations. It is important to note that an increase of one pH point results in a ten-fold decrease in hydrogen ion concentration.

Table 1: pH and Hydrogen ion concentration

pH	*[H⁺] nanomol/l*
6.0	1000
7.0	100
8.0	10
9.0	1

Teaching Point

As a solution becomes more acidic or less alkaline, the pH falls (hydrogen ion concentration rises). The opposite happens when solutions become less acidic or more alkaline.

Although pH terminology is widely used in textbooks and in biochemistry reports, it is important to realise that pH is merely a reflection of the hydrogen ion concentration. In the rest of this article both terms will be used to impress upon the reader that, essentially, they refer to the same thing.

Acids, Bases and Buffers

Acids: An acid is defined as any compound, which forms hydrogen ions in solution. For this reason acids are sometimes referred to as "proton donors". To aid understanding of these concepts consider an imaginary acid with the chemical formula HA.

Bases: A base is a compound that combines with hydrogen ions in solution. Therefore, bases can be referred to as "proton acceptors".

Strong Acids: A strong acid is a compound that ionizes completely in solution to form hydrogen ions and a base. Example 2 illustrates a strong acid in solution, where this dissociation is complete.

Weak Acids and Bases: these are compounds that are only partially ionised in solution.

Buffers: A buffer is a compound that limits the change in hydrogen ion concentration (and so pH) when hydrogen ions are added or removed from the solution. It may be useful to think of the buffer as being like a sponge. When hydrogen ions are in excess, the sponge mops up the extra ions. When in short supply the sponge can be squeezed out to release more hydrogen ions! All buffers are in fact weak acids or bases. This creates more HA whilst removing the excess H^+ from the solution. Similarly, as hydrogen ions are removed from solution by addition of a strong base the reaction moves to the right restoring the hydrogen ion concentration and reducing the quantity of HA. The effects of buffers can also be illustrated graphically. Notice that during the highlighted portion of the curve a large volume of acid is added with little change in [H^+] or pH. As we shall see later buffers are crucial in maintaining hydrogen ions within a narrow range concentrations in the body.

The Importance of Hydrogen Ion Concentration

Hydrogen ion concentration has a widespread effect on the function of the body's enzyme systems. The hydrogen ion is highly reactive and will combine with bases or negatively charged ions at very low concentrations. Proteins contain many negatively charged and basic groups within their structure. Thus, a change in pH will alter the degree ionization of a protein, which may in turn affect its functioning. At more extreme hydrogen ion concentrations a protein's structure may be completely disrupted (the protein is then said to be denatured). Although most enzymes function optimally around physiological pH it should be noted that a few enzymes function best at a much higher hydrogen ion concentration (ie: at a lower pH). The most notable of these enzymes is pepsin, which works best in the acid environment of the stomach -optimum pH 1.5-3 or [H^+]= 3-30 million nanomol/l.

As enzymes have a huge number of functions around the body, an abnormal pH can result in disturbances in a wide range of body systems. Thus, disturbances in pH may result in abnormal respiratory and cardiac function, derangements in blood clotting and drug metabolism, to name but a few. From these few examples it is clear that the anaesthetist should strive to ensure that hydrogen ion concentration is maintained within the normal range.

Biological and Chemical Organization of Cell

The cell is the functional basic unit of life. It was discovered by Robert Hooke and is the functional unit of all known living organisms. It is the smallest unit of life that is classified as a living thing, and is often called the building block of life. Some organisms, such as most bacteria, are unicellular (consist of a single cell). Other organisms, such as humans, are multicellular. Humans have about 100 trillion or 10^{14} cells; a typical cell size is 10 μm and a typical cell mass is 1 nanogram. The largest cells are about 135 μm in the anterior horn in the spinal cord while granule cells in the cerebellum, the smallest, can be some 4 μm and the longest cell can reach from the toe to the lower brain stem (Pseudounipolar cells). The largest known cells are unfertilised ostrich egg cells which weigh 3.3 pounds.

In 1835, before the final cell theory was developed, Jan Evangelista Purkynì observed small "granules" while looking at the plant tissue through a microscope. The cell theory, first developed in 1839 by Matthias Jakob Schleiden and Theodor Schwann, states that all organisms are composed of one or more cells, that all cells come from preexisting cells, that vital functions of an organism occur within cells, and that all cells contain the hereditary information necessary for regulating cell functions and for transmitting information to the next generation of cells.

The word *cell* comes from the Latin *cellula*, meaning, a small room. The descriptive term for the smallest living biological structure was coined by Robert Hooke in a book he published in 1665 when he compared the cork cells he saw through his microscope to the small rooms monks lived in.

Anatomy

There are two types of cells: eukaryotic and prokaryotic. Prokaryotic cells are usually independent, while eukaryotic cells are often found in multicellular organisms.

Prokaryotic Cells

The prokaryote cell is simpler, and therefore smaller, than a eukaryote cell, lacking a nucleus and most of the other organelles of eukaryotes. There are two kinds of prokaryotes: bacteria and archaea; these share a similar structure.

Nuclear material of prokaryotic cell consist of a single chromosome which is in direct contact with cytoplasm. Here the undefined nuclear region in the cytoplasm is called nucleoid.

A prokaryotic cell has three architectural regions:

- On the outside, flagella and pili project from the cell's surface. These are structures (not present in all prokaryotes) made of proteins that facilitate movement and communication between cells;
- Enclosing the cell is the cell envelope – generally consisting of a cell wall covering a plasma membrane though some bacteria also have a further covering layer called a capsule. The envelope gives rigidity to the cell and separates the interior of the cell from its environment, serving as a protective filter. Though most prokaryotes have a cell wall, there are exceptions such as *Mycoplasma* (bacteria) and *Thermoplasma* (archaea). The cell wall consists of *peptidoglycan* in bacteria, and acts as an additional barrier against exterior forces. It also prevents the cell from expanding and finally bursting (cytolysis) from osmotic pressure against a hypotonic environment. Some eukaryote cells (plant cells and fungi cells) also have a cell wall;
- Inside the cell is the cytoplasmic region that contains the cell genome (DNA) and ribosomes and various sorts of inclusions. A

prokaryotic chromosome is usually a circular molecule (an exception is that of the bacterium *Borrelia burgdorferi*, which causes Lyme disease). Though not forming a *nucleus*, the DNA is condensed in a *nucleoid*. Prokaryotes can carry extrachromosomal DNA elements called *plasmids*, which are usually circular. Plasmids enable additional functions, such as antibiotic resistance.

Eukaryotic Cells

Eukaryotic cells are about 15 times wider than a typical prokaryote and can be as much as 1000 times greater in volume. The major difference between prokaryotes and eukaryotes is that eukaryotic cells contain membrane-bound compartments in which specific metabolic activities take place. Most important among these is a cell nucleus, a membrane-delineated compartment that houses the eukaryotic cell's DNA. This nucleus gives the eukaryote its name, which means "true nucleus." Other differences include:

- The plasma membrane resembles that of prokaryotes in function, with minor differences in the setup. Cell walls may or may not be present.
- The eukaryotic DNA is organized in one or more linear molecules, called chromosomes, which are associated with histone proteins. All chromosomal DNA is stored in the *cell nucleus*, separated from the cytoplasm by a membrane. Some eukaryotic organelles such as mitochondria also contain some DNA.
- Many eukaryotic cells are ciliated with *primary cilia*. Primary cilia play important roles in chemosensation, mechanosensation, and thermosensation. Cilia may thus be "viewed as sensory cellular antennae that coordinate a large number of cellular signalling pathways, sometimes coupling the signalling to ciliary motility or alternatively to cell division and differentiation."
- Eukaryotes can move using *motile cilia* or *flagella*. The flagella are more complex than those of prokaryotes.

Subcellular Components

All cells, whether prokaryotic or eukaryotic, have a membrane that envelops the cell, separates its interior from its environment, regulates what moves in and out (selectively permeable), and maintains the electric potential of the cell. Inside the membrane, a salty cytoplasm takes up most of the cell volume.

All cells possess DNA, the hereditary material of genes, and RNA, containing the information necessary to build various proteins such as enzymes, the cell's primary machinery. There are also other kinds of biomolecules in cells. This article will list these primary components of the cell, then briefly describe their function.

Membrane

The cell membrane is a biological membrane that separates the interior of all cells from the outside environment.. The cell membrane is selectively-permeable to ions and organic molecules and controls the movement of substances in and out of cells.. It consists of the phospholipid bilayer, consisting of hydrophobic tails and hydrophilic heads with embedded proteins, which are involved in a variety of cellular processes such as cell adhesion, ion conductivity and cell signalling. The plasma membrane also serves as the attachment surface for the extracellular glycocalyx and cell wall and intracellular cytoskeleton.

Function

The cell membrane surrounds the protoplasm of a cell and, in animal cells, physically separates the intracellular components from the extracellular environment. Fungi, bacteria and plants also have the cell wall which provides a mechanical support for the cell and precludes passage of the larger molecules. The cell membrane also plays a role in anchoring the cytoskeleton to provide shape to the cell, and in attaching to the extracellular matrix and other cells to help group cells together to form tissues. The barrier is differentially permeable and able to regulate what enters and exits the cell, thus facilitating the transport of materials needed for survival. The movement of substances across the membrane can be either *passive,* occurring without the input of cellular energy, or active, requiring the cell to expend energy in moving it. The membrane also maintains the cell potential.

Structure

Fluid Mosaic Model: According to the fluid mosaic model of S. J. Singer and Garth Nicolson 1972, the biological membranes can be considered as a two-dimensional liquid where all lipid and protein molecules diffuse more or less easily. This picture may be valid in the space scale of 10 nm. However, the plasma membranes contain different structures or domains that can be classified as: (a) protein-protein complexes; (b) lipid rafts, and (c) pickets and fences formed by the actin-based cytoskeleton.

Lipid Bilayer

The cell membrane consists primarily of a thin layer of amphipathic phospholipids which spontaneously arrange so that the hydrophobic "tail" regions are shielded from the surrounding polar fluid, causing the more hydrophilic "head" regions to associate with the cytosolic and extracellular faces of the resulting bilayer. This forms a continuous, spherical lipid bilayer. The arrangement of hydrophilic heads and hydrophobic tails of the lipid bilayer prevent polar solutes (e.g. amino acids, nucleic acids, carbohydrates, proteins, and ions) from diffusing across the membrane, but generally allows for the passive diffusion of hydrophobic molecules. This affords the cell the ability to control the movement of these substances via transmembrane protein complexes such as pores and gates.

Flippases and Scramblases concentrate phosphatidyl serine, which carries a negative charge, on the inner membrane. Along with NANA, this creates an extra barrier to charged moieties moving through the membrane.

Membranes serve diverse functions in eukaryotic and prokaryotic cells. One important role is to regulate the movement of materials into and out of cells. The phospholipid bilayer structure (fluid mosaic model) with specific membrane proteins accounts for the selective permeability of the membrane and passive and active transport mechanisms. In addition, membranes in prokaryotes and in the mitochondria and chloroplasts of eukaryotes facilitate the synthesis of ATP through chemiosmosis.

Membrane Polarity

The apical membrane of a polarized cell is the surface of the plasma membrane that faces the lumen. This is particularly evident in epithelial and endothelial cells, but also describes other polarized cells, such as neurons. The basolateral membrane of a polarized cell is the surface of the plasma membrane that forms its basal and lateral surfaces. It faces towards the interstitium, and away from the lumen.

"Basolateral membrane" is a compound phrase referring to the terms *basal (base) membrane* and *lateral (side) membrane*, which, especially in epithelial cells, are identical in composition and activity.

Proteins (such as ion channels and pumps) are free to move from the basal to the lateral surface of the cell or *vice versa* in accordance with the fluid mosaic model. Tight junctions that join epithelial cells near their apical surface prevent the migration of proteins from the basolateral membrane to the apical membrane. The basal and lateral surfaces thus remain roughly equivalent to one another, yet distinct from the apical surface.

Integral Membrane Proteins

The cell membrane contains many integral membrane proteins, which pepper the entire surface. These structures, which can be visualized by electron microscopy or fluorescence microscopy, can be found on the inside of the membrane, the outside, or membrane spanning. These may include integrins, cadherins, desmosomes, clathrin-coated pits, caveolaes, and different structures involved in cell adhesion.

Membrane Skeleton

The cytoskeleton is found underlying the cell membrane in the cytoplasm and provides a scaffolding for membrane proteins to anchor to, as well as forming organelles that extend from the cell. Indeed, cytoskeletal elements interact extensively and intimately with the cell membrane. Anchoring proteins restricts them to a particular cell surface — for example, the *apical surface* of epithelial cells that line the vertebrate gut — and limits how far they may diffuse within the bilayer. The cytoskeleton is able to form appendage-like organelles, such as cilia, which are microtubule-based extensions covered by the cell membrane, and filopodia, which are actin-based extensions. These extensions are ensheathed in membrane and project from the surface of the cell in order to sense the external environment and/or make contact with the substrate or other cells.

The apical surfaces of epithelial cells are dense with actin-based finger-like projections known as microvilli, which increase cell surface area and thereby increase the absorption rate of nutrients. Localized decoupling of the cytoskeleton and cell membrane results in formation of a bleb.

Composition

Cell membranes contain a variety of biological molecules, notably lipids and proteins. Material is incorporated into the membrane, or deleted from it, by a variety of mechanisms:

- Fusion of intracellular vesicles with the membrane (exocytosis) not only excretes the contents of the vesicle but also incorporates the vesicle membrane's components into the cell membrane. The membrane may form blebs around extracellular material that pinch off to become vesicles (endocytosis).
- If a membrane is continuous with a tubular structure made of membrane material, then material from the tube can be drawn into the membrane continuously.
- Although the concentration of membrane components in the aqueous phase is low (stable membrane components have low solubility in water), there is an exchange of molecules between the lipid and aqueous phases.

Lipids

The cell membrane consists of three classes of amphipathic lipids: phospholipids, glycolipids, and cholesterols. The amount of each depends upon the type of cell, but in the majority of cases phospholipids are the most abundant. In RBC studies, 30% of the plasma membrane is lipid.

The fatty chains in phospholipids and glycolipids usually contain an even number of carbon atoms, typically between 16 and 20. The 16- and 18-carbon fatty acids are the most common.

Fatty acids may be saturated or unsaturated, with the configuration of the double bonds nearly always *cis*. The length and the degree of unsaturation of fatty acid chains have a profound effect on membrane fluidity as unsaturated lipids create a kink, preventing the fatty acids from packing together as tightly, thus decreasing the melting temperature (increasing the fluidity) of the membrane. The ability of some organisms to regulate the fluidity of their cell membranes by altering lipid composition is called homeoviscous adaptation.

The entire membrane is held together via non-covalent interaction of hydrophobic tails, however the structure is quite fluid and not fixed rigidly in place. Under physiological conditions phospholipid molecules in the cell membrane are in the liquid crystalline state.

It means the lipid molecules are free to diffuse and exhibit rapid lateral diffusion along the layer in which they are present. However, the exchange of phospholipid molecules between intracellular and extracellular leaflets of the bilayer is a very slow process. Lipid rafts and caveolae are examples of cholesterol-enriched microdomains in the cell membrane.

In animal cells cholesterol is normally found dispersed in varying degrees throughout cell membranes, in the irregular spaces between the hydrophobic tails of the membrane lipids, where it confers a stiffening and strengthening effect on the membrane.

Carbohydrates

Plasma membranes also contain carbohydrates, predominantly glycoproteins, but with some glycolipids (cerebrosides and gangliosides). For the most part, no glycosylation occurs on membranes within the cell; rather generally glycosylation occurs on the extracellular surface of the plasma membrane.

The glycocalyx is an important feature in all cells, especially epithelia with microvilli. Recent data suggest the glycocalyx participates in cell adhesion, lymphocyte homing, and many others.

The penultimate sugar is galactose and the terminal sugar is sialic acid, as the sugar backbone is modified in the golgi apparatus. Sialic acid carries a negative charge, providing an external barrier to charged particles.

Proteins

Type	*Description*	*Examples*
Integral proteins or *transmembrane proteins*	Span the membrane and have a hydrophilic cytosolic domain, which interacts with internal molecules, a hydrophobic membrane spanning domain that anchors it within the cell membrane, and a hydrophilic extracellular domain that interacts with external molecules. The hydrophobic domain consists of one, multiple, or a combination of α-helices and β sheet protein motifs.	Ion channels, proton pumps, G protein-coupled receptor
Lipid anchored proteins	Covalently-bound to single or multiple lipid molecules; hydrophobically insert into the cell membrane and anchor the protein. The protein itself is not in contact with the membrane.	G proteins
Peripheral proteins	Attached to integral membranc proteins, or associated with peripheral regions of the lipid bilayer. These proteins tend to have only temporary interactions with biological membranes, and, once reacted the molecule, dissociates to carry on its work in the cytoplasm.	Some enzymes, some hormones

The cell membrane plays host to a large amount of protein that is responsible for its various activities. The amount of protein differs between species and according to function, however the typical amount in a cell membrane is 50%. These proteins are undoubtedly important to a cell: Approximately a third of the genes in yeast code specifically for them, and this number is even higher in multicellular organisms. The cell membrane, being exposed to the outside environment, is an important site of cell-cell communication. As such, a large variety of protein receptors and identification proteins, such as antigens, are present on the surface of the membrane. Functions of membrane proteins can also include cell-cell contact, surface recognition, cytoskeleton contact, signalling, enzymatic activity, or transporting substances across the membrane.

Most membrane proteins must be inserted in some way into the membrane. For this to occur, an N-terminus "signal sequence" of amino acids directs proteins to the endoplasmic reticulum, which inserts the proteins into a lipid bilayer. Once inserted, the proteins are then transported to their final destination in vesicles, where the vesicle fuses with the target membrane.

Variation

The cell membrane has different lipid and protein compositions in distinct types of cells and may have therefore specific names for certain cell types:

- Sarcolemma in myocytes
- Oolemma in oocytes.

Permeability

The permeability of a membrane is the ease of molecules to pass through it. Permeability depends mainly on the electric charge of the molecule and to a lesser extent the molar mass of the molecule. Electrically neutral and small molecules pass the membrane easier than charged, large ones. The inability of charged molecules to pass through the cell membrane results in pH parturition of substances throughout the fluid compartments of the body.

Cytoskeleton

The cytoskeleton (also CSK) is a cellular "scaffolding" or "skeleton" contained within the cytoplasm and is made out of protein. The cytoskeleton is present in all cells; it was once thought to be unique to eukaryotes, but recent research has identified the prokaryotic cytoskeleton. It has structures such as flagella, cilia and lamellipodia and plays important roles in both intracellular transport (the movement of vesicles and organelles, for example) and cellular division. The concept of a protein mosaic that dynamically coordinated cytoplasmic biochemistry was proposed by Rudolph Peters in 1929 while the term (*cytosquelette*, in French) was first introduced by French embryologist Paul Wintrebert in 1931. It is almost like the human skeleton.

The Eukaryotic Cytoskeleton

Eukaryotic cells contain three main kinds of cytoskeletal filaments, which are microfilaments, intermediate filaments, and microtubules. The cytoskeleton provides the cell with structure and shape, and by excluding macromolecules from some of the cytosol it adds to the level of macromolecular crowding in this compartment. Cytoskeletal elements interact extensively and intimately with cellular membranes.

Intermediate Filaments

These filaments, around 10 manometers in diameter, are more stable (strongly bound) than actin filaments, and heterogeneous constituents of the cytoskeleton. Although little work has been done on intermediate filaments in plants, there is some evidence that cytosolic intermediate filaments might be present, and plant nuclear filaments have been detected. Like actin filaments, they function in the maintenance of cell-shape by bearing tension (microtubules, by contrast, resist compression. It may be useful to think of micro- and intermediate filaments as cables, and of microtubules as cellular support beams). Intermediate filaments organize the internal tridimensional structure of the cell, anchoring organelles and serving as structural components of the nuclear lamina and sarcomeres. They also participate in some cell-cell and cell-matrix junctions.

Different intermediate filaments are:

- made of vimentins, being the common structural support of many cells.
- made of keratin, found in skin cells, hair and nails.
- neurofilaments of neural cells.
- made of lamin, giving structural support to the nuclear envelope.

Microtubules

Microtubules are hollow cylinders about 23 nm in diameter (lumen = approximately 15nm in diameter), most commonly comprising 13 protofilaments which, in turn, are polymers of alpha and beta tubulin. They have a very dynamic behaviour, binding GTP for polymerization. They are commonly organized by the centrosome.

In nine triplet sets (star-shaped), they form the centrioles, and in nine doublets oriented about two additional microtubules (wheel-shaped) they form cilia and flagella. The latter formation is commonly referred to as a "9+2" arrangement, wherein each doublet is connected to another by the protein dynein. As both flagella and cilia are structural components of the cell, and are maintained by microtubules, they can be considered part of the cytoskeleton.

They play key roles in:

- intracellular transport (associated with dyneins and kinesins, they transport organelles like mitochondria or vesicles).
- the axoneme of cilia and flagella.
- the mitotic spindle.
- synthesis of the cell wall in plants.

The Prokaryotic Cytoskeleton

The cytoskeleton was previously thought to be a feature only of eukaryotic cells, but homologues to all the major proteins of the eukaryotic cytoskeleton have recently been found in prokaryotes.

Although the evolutionary relationships are so distant that they are not obvious from protein sequence comparisons alone, the similarity of their three-dimensional structures and similar functions in maintaining cell shape and polarity provides strong evidence that the eukaryotic and prokaryotic cytoskeletons are truly homologous. However, some structures in the bacterial cytoskeleton may have yet to be identified.

FtsZ

FtsZ was the first protein of the prokaryotic cytoskeleton to be identified. Like tubulin, FtsZ forms filaments in the presence of GTP, but these filaments do not group into tubules. During cell division, FtsZ is the first protein to move to the division site, and is essential for recruiting other proteins that synthesize the new cell wall between the dividing cells.

MreB and ParM

Prokaryotic actin-like proteins, such as MreB, are involved in the maintenance of cell shape. All non-spherical bacteria have genes encoding actin-like proteins, and these proteins form a helical network beneath the cell membrane that guides the proteins involved in cell wall biosynthesis.

Some plasmids encode a partitioning system that involves an actin-like protein ParM. Filaments of ParM exhibit dynamic instability, and may partition plasmid DNA into the dividing daughter cells by a mechanism analogous to that used by microtubules during eukaryotic mitosis.

Crescentin

The bacterium *Caulobacter crescentus* contains a third 3rd protein, crescentin, that is related to the intermediate filaments of eukaryotic cells. Crescentin is also involved in maintaining cell shape, such as helical and vibrioid forms of bacteria, but the mechanism by which it does this is currently unclear.

History

Microtrabeculae

A fourth eukaryotic cytoskeletal element, *microtrabeculae*, was proposed by Keith Porter based on images obtained from high-voltage electron microscopy of whole cells in the 1970s. Porter proposed that this microtrabecular structure represented a novel filamentous network distinct from microtubules, filamentous actin, or intermediate filaments. It is now generally accepted that microtrabeculae are nothing more than an artifact of certain types of fixation treatment, although we have yet to fully understand the complexity of the cell's cytoskeleton.

The cytoskeleton acts to organize and maintain the cell's shape; anchors organelles in place; helps during endocytosis, the uptake of external materials by a cell, and cytokinesis, the separation of daughter cells after cell division; and moves parts of the cell in processes of growth and mobility. The eukaryotic cytoskeleton is composed of microfilaments, intermediate filaments and microtubules. There is a great number of proteins associated with them, each controlling a cell's structure by directing, bundling, and aligning filaments. The prokaryotic cytoskeleton is less well-studied but is involved in the maintenance of cell shape, polarity and cytokinesis.

Organelles

The human body contains many different organs, such as the heart, lung, and kidney, with each organ performing a different function. Cells also have a set of "little organs," called organelles, that are adapted and/or specialized for carrying out one or more vital functions. Both eukaryotic and prokaryotic cells have organelles but organelles in eukaryotes are generally more complex and may be membrane bound. There are several types of organelles in a cell. Some (such as the nucleus and golgi apparatus) are typically solitary, while others (such as mitochondria, peroxisomes and lysosomes) can be numerous (hundreds to thousands). The cytosol is the gelatinous fluid that fills the cell and surrounds the organelles.

Structures Outside the Cell Wall

Capsule

A gelatinous capsule is present in some bacteria outside the cell wall. The capsule may be polysaccharide as in pneumococci, meningococci or polypeptide as *Bacillus anthracis* or hyaluronic acid as in streptococci. Capsules are not marked by ordinary stain and can be detected by special stain. The capsule is antigenic. The capsule has antiphagocytic function so it determines the virulence of many bacteria. It also plays a role in attachment of the organism to mucous membranes.

Flagella

Flagella are the organelles of cellular mobility. They arise from cytoplasm and extrude through the cell wall. They are long and thick thread-like appendages, protein in nature. Are most commonly found in bacteria cells but are found in animal cells as well.

Fimbriae (Pili)

They are short and thin hair like filaments, formed of protein called pilin (antigenic). Fimbriae are responsible for attachment of bacteria to specific receptors of human cell (adherence). There are special types of pili called (sex pili) involved in conjunction.

Functions

Growth and Metabolism

Between successive cell divisions, cells grow through the functioning of cellular metabolism. Cell metabolism is the process by which individual cells process nutrient molecules. Metabolism has two distinct divisions: catabolism, in which the cell breaks down complex molecules to produce energy and reducing power, and anabolism, in which the cell uses energy and reducing power to construct complex molecules and perform other biological functions. Complex sugars consumed by the organism can be broken down into a less chemically complex sugar molecule called glucose. Once inside the cell, glucose is broken down to make adenosine triphosphate (ATP), a form of energy, through two different pathways.

The first pathway, glycolysis, requires no oxygen and is referred to as anaerobic metabolism. Each reaction is designed to produce some hydrogen ions that can then be used to make energy packets (ATP). In prokaryotes, glycolysis is the only method used for converting energy.

The second pathway, called the Krebs cycle, or citric acid cycle, occurs inside the mitochondria and can generate enough ATP to run all the cell functions.

Creation

Cell division involves a single cell (called a *mother cell*) dividing into two daughter cells. This leads to growth in multicellular organisms (the growth of tissue) and to procreation (vegetative reproduction) in unicellular organisms.

Prokaryotic cells divide by binary fission. Eukaryotic cells usually undergo a process of nuclear division, called mitosis, followed by division of the cell, called cytokinesis. A diploid cell may also undergo meiosis to produce haploid cells, usually four. Haploid cells serve as gametes in multicellular organisms, fusing to form new diploid cells.

DNA replication, or the process of duplicating a cell's genome, is required every time a cell divides. Replication, like all cellular activities, requires specialized proteins for carrying out the job.

Protein Synthesis

Cells are capable of synthesizing new proteins, which are essential for the modulation and maintenance of cellular activities. This process involves the formation of new protein molecules from amino acid building blocks based on information encoded in DNA/RNA. Protein synthesis generally consists of two major steps: transcription and translation. Transcription is the process where genetic information in DNA is used to produce a complementary RNA strand. This RNA strand is then processed to give messenger RNA (mRNA), which is free to migrate through the cell. mRNA molecules bind to protein-RNA complexes called ribosomes located in the cytosol, where they are translated into polypeptide sequences. The ribosome mediates the formation of a polypeptide sequence based on the mRNA sequence. The mRNA sequence directly relates to the polypeptide sequence by binding to transfer RNA (tRNA) adapter molecules in binding pockets within the ribosome. The new polypeptide then folds into a functional three-dimensional protein molecule.

Movement or Motility

Cells can move during many processes: such as wound healing, the immune response and cancer metastasis. For wound healing to occur, white blood cells and cells that ingest bacteria move to the wound site to kill the microorganisms that cause infection. At the same time fibroblasts (connective tissue cells) move there to remodel damaged structures. In the case of tumour development, cells from a primary tumour move away and spread to other parts of the body. Cell motility involves many receptors, crosslinking, bundling, binding, adhesion, motor and other proteins. The process is divided into three steps – protrusion of the leading edge of the cell, adhesion of the leading edge and de-adhesion at the cell body and rear, and cytoskeletal contraction to pull the cell forward. Each step is driven by physical forces generated by unique segments of the cytoskeleton.

Evolution

The origin of cells has to do with the origin of life, which began the history of life on Earth.

Origin of the First Cell

There are several theories about the origin of small molecules that could lead to life in an early Earth. One is that they came from meteorites. Another is that they were created at deep-sea vents. A third is that they were synthesized by lightning in a reducing atmosphere; although it is not clear if Earth had such an atmosphere. There are essentially no experimental data defining what the first self-replicating forms were. RNA is generally assumed to be the earliest self-replicating molecule, as it is capable of both storing genetic information and catalysing chemical reactions. But some other entity with the potential to self-replicate could have preceded RNA, like clay or peptide nucleic acid.

Cells emerged at least 4.0–4.3 billion years ago. The current belief is that these cells were heterotrophs. An important characteristic of cells is the cell membrane, composed of a bilayer of lipids. The early cell membranes were probably more simple and permeable than modern ones, with only a single fatty acid chain per lipid. Lipids are known to spontaneously form bilayered vesicles in water, and could have preceded RNA. But the first cell membranes could also have been produced by catalytic RNA, or even have required structural proteins before they could form.

Origin of Eukaryotic Cells

The eukaryotic cell seems to have evolved from a symbiotic community of prokaryotic cells. DNA-bearing organelles like the mitochondria and the chloroplasts are almost certainly what remains of ancient symbiotic oxygen-breathing proteobacteria and cyanobacteria, respectively, where the rest of the cell seems to be derived from an ancestral archaean prokaryote cell – a theory termed the endosymbiotic theory. There is still considerable debate about whether organelles like the hydrogenosome predated the origin of mitochondria.

Sex, as the stereotyped choreography of meiosis and syngamy that persists in nearly all extant eukaryotes, may have played a role in the transition from prokaryotes to eukaryotes. An 'origin of sex as vaccination' theory suggests that the eukaryote genome accreted from prokaryan parasite genomes in numerous rounds of lateral gene transfer. Sex-as-syngamy (fusion sex) arose when infected hosts began swapping nuclearized genomes containing co-evolved, vertically transmitted symbionts that conveyed protection against horizontal infection by more virulent symbionts.

History

- 1632–1723: Antonie van Leeuwenhoek teaches himself to grind lenses, builds a microscope and draws protozoa, such as *Vorticella* from rain water, and bacteria from his own mouth.
- 1665: Robert Hooke discovers cells in cork, then in living plant tissue using an early microscope.
- 1839: Theodor Schwann and Matthias Jakob Schleiden elucidate the principle that plants and animals are made of cells, concluding that cells are a common unit of structure and development, and thus founding the cell theory.
- The belief that life forms can occur spontaneously (*generatio spontanea*) is contradicted by Louis Pasteur (1822–1895) (although Francesco Redi had performed an experiment in 1668 that suggested the same conclusion).
- 1855: Rudolph Virchow states that cells always emerge from cell divisions (*omnis cellula ex cellula*).
- 1931: Ernst Ruska builds first transmission electron microscope (TEM) at the University of Berlin. By 1935, he has built an EM with twice the resolution of a light microscope, revealing previously unresolvable organelles.
- 1953: Watson and Crick made their first announcement on the double-helix structure for DNA on February 28.
- 1981: Lynn Margulis published *Symbiosis in Cell Evolution* detailing the endosymbiotic theory.

2

Molecules Biochemistry

Most of the Universe consists of matter and energy. Energy is the capacity to do work. Matter has mass and occupies space. All matter is composed of basic elements that cannot be broken down to substances with different chemical or physical properties. Elements are substances consisting of one type of atom, for example Carbon atoms make up diamond, and also graphite. Pure (24K) gold is composed of only one type of atom, gold atoms. Atoms are the smallest particle into which an element can be divided. The ancient Greek philosophers developed the concept of the atom, although they considered it the fundamental particle that could not be broken down. Since the work of Enrico Fermi and his colleagues, we now know that the atom is divisible, often releasing tremendous energies as in nuclear explosions or (in a controlled fashion in) thermonuclear power plants.

Subatomic particles were discovered during the 1800s. For our purposes we will concentrate only on three of them, summarized in Table 1. The proton is located in the centre (or nucleus) of an atom, each atom has at least one proton. Protons have a charge of +1, and a mass of approximately 1 atomic mass unit (amu). Elements differ from each other in the number of protons they have, e.g. Hydrogen has 1 proton; Helium has 2.

The neutron also is located in the atomic nucleus (except in Hydrogen). The neutron has no charge, and a mass of slightly over 1 amu. Some scientists propose the neutron is made up of a proton and electron-like particle.

The electron is a very small particle located outside the nucleus. Because they move at speeds near the speed of light the precise location of electrons is hard to pin down. Electrons occupy orbitals, or areas where they have a high statistical probability of occurring. The charge on an electron is -1. Its mass is negligible (approximately 1800 electrons are needed to equal the mass of one proton).

Table 1. Subatomic particles of use in biology.

Name	*Charge*	*Location*	*Mass*
Proton	+1	atomic nucleus	1.6726×10^{-27} kg
Neutron	0	atomic nucleus	1.6750×10^{-27} kg
Electron	-1	electron orbital	9.1095×10^{-31} kg

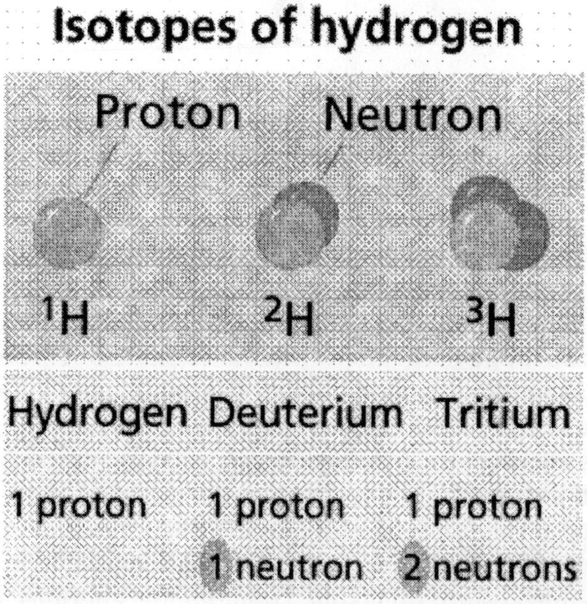

The atomic number is the number of protons an atom has. It is characteristic and unique for each element. The atomic mass (also referred to as the atomic weight) is the number of protons and neutrons in an atom. Atoms of an element that have differing numbers of neutrons (but a constant atomic number) are termed isotopes. Isotopes, can be used to determine the diet of ancient peoples by determining proportions of isotopes in mummified or fossilized human tissues. Biochemical pathways can be deciphered by using isotopic tracers. The age of fossils and artifacts can be determined by using radioactive isotopes, either directly on the fossil or on the rocks that surround the fossil. Isotopes are also the source of radiation used in medical diagnostic and treatment procedures.

Note that each of these isotopes of hydrogen has only one proton. Isotopes differ from each other in the number of neutrons, not in the number of protons.

Some isotopes are radioisotopes, which spontaneously decay, releasing radioactivity. Other isotopes are stable. Examples of radioisotopes are Carbon-14 (symbol ^{14}C), and deuterium (also known as Hydrogen-2; ^{2}H). Stable isotopes are ^{12}C and ^{1}H.

Carbon has three isotopes, of which carbon-12 and carbon-14 are the most well known

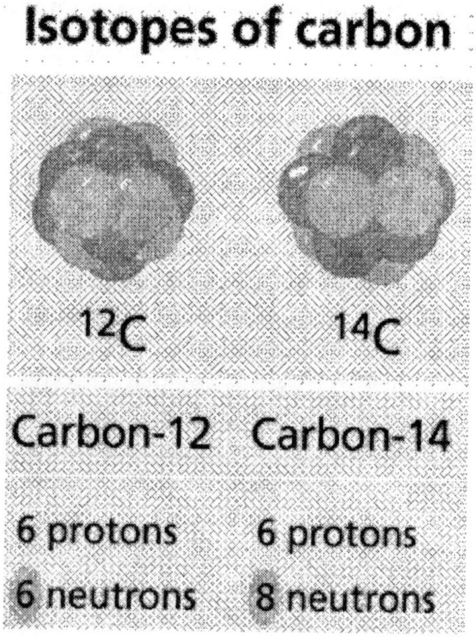

Electrons and Energy

Electrons, because they move so fast (approximately at the speed of light), seem to straddle the fence separating energy from matter. Albert Einstein developed his famous $E=mc^2$ equation relating matter and energy over a century ago. Because of his (and others) work, we think of electrons both as particles of matter (having mass is a property of matter) and as units (or quanta) of energy.

Excitation of an electron by energy, causing the electron to "jump" to another electron (energy) level known as the excited state.

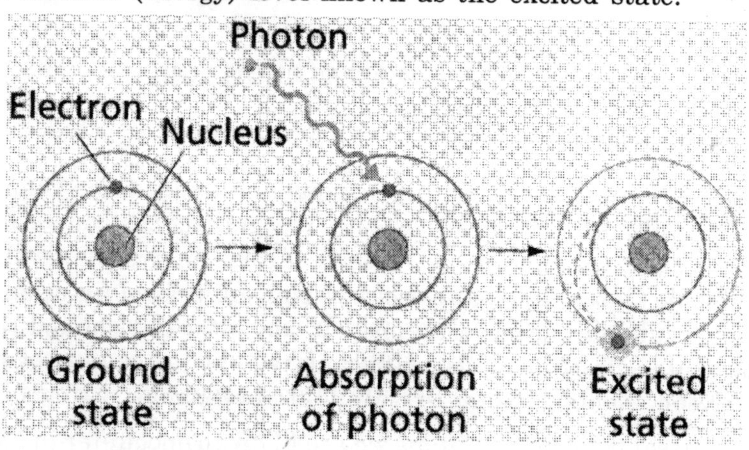

An orbital is also an area of space in which an electron will be found 90% of the time. Orbitals have a variety of shapes. Each orbital has a characteristic energy state and a characteristic shape.

The *s* orbital is spherical. Since each orbital can hold a maximum of two electrons, atomic numbers above 2 must fill the other orbitals. The p_x, p_y, and p_z orbitals are dumbbell shaped, along the x, y, and z axes respectively.

Energy levels are located a certain "distance" from the nucleus. The major energy levels into which electrons fit, are (from the nucleus outward) K, L, M, and N. Sometimes these are numbered, with electron configurations being: $1s^2 2s^2 2p^1$, (where the first shell K is indicated with the number 1, the second shell L with the number 2, etc.). This nomenclature tells us that for the atom mentioned in this paragraph, the first energy level (shell) has two electrons in its *s* orbital (the only orbital it can have), and second energy level has a maximum of two electrons in its *s* orbital, plus one electron in its *p* orbital.

Geometry of orbitals. S-orbitals are spherical, p-orbitals are shaped like a dumbbell

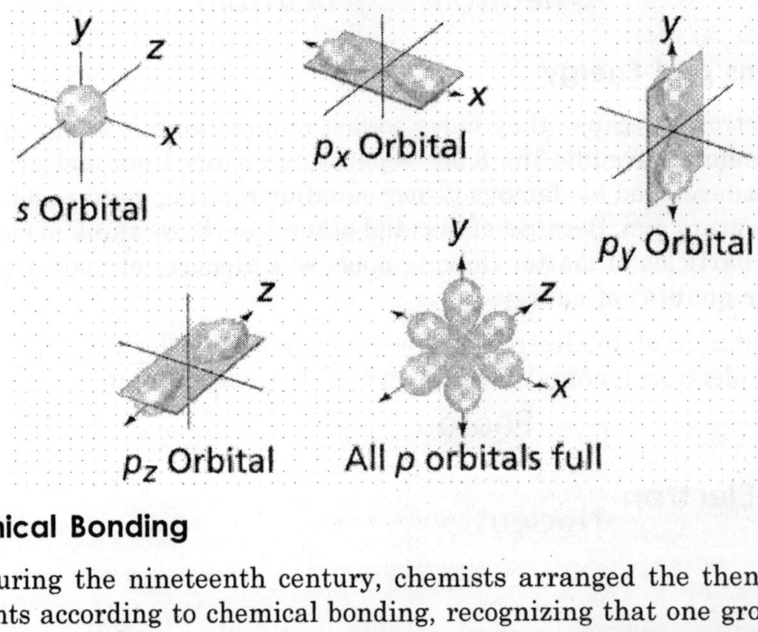

Chemical Bonding

During the nineteenth century, chemists arranged the then-known elements according to chemical bonding, recognizing that one group (the furthermost right column on the Periodic Table, referred to as the Inert Gases or Noble Gases) tended to occur in elemental form (in other words, not in a molecule with other elements). It was later determined that this group had outer electron shells containing two (as in the case of Helium) or eight (Neon, Xenon, Radon, Krypton, etc.) electrons.

As a general rule, for the atoms we are likely to encounter in biological systems, atoms tend to gain or lose their outer electrons to achieve a Noble Gas outer electron shell configuration of two or eight electrons. The number of electrons that are gained or lost is characteristic for each element, and ultimately determines the number and types of chemical bonds atoms of that element can form.

Atomic diagrams illustrating the filling of the outer electron shells.

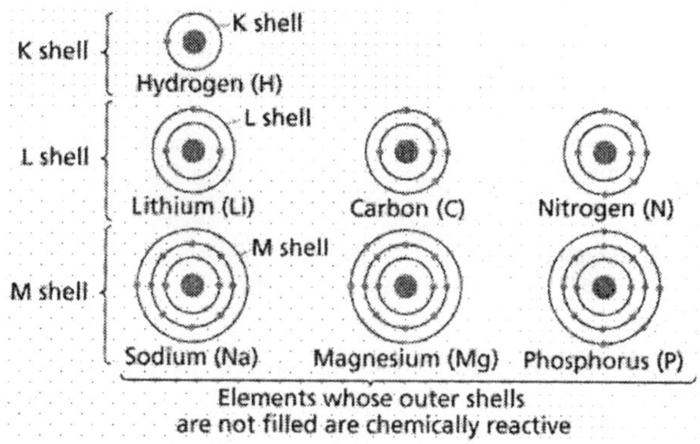

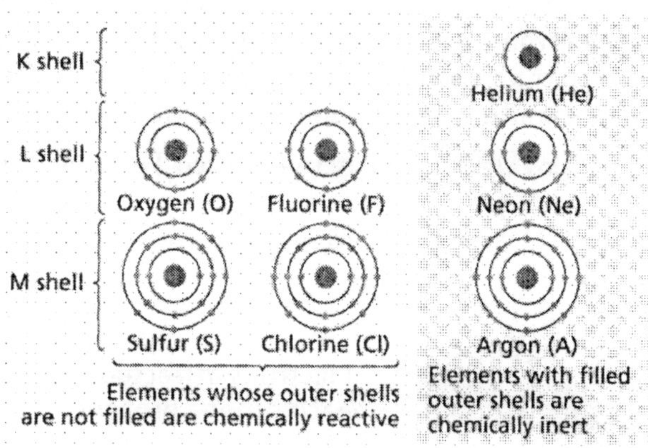

Ionic bonds are formed when atoms become ions by gaining or losing electrons. Chlorine is in a group of elements having seven electrons in their outer shells. Members of this group tend to gain one electron, acquiring a charge of -1. Sodium is in another group with elements having one electron in their outer shells. Members of this group tend to lose that outer electron, acquiring a charge of +1. Oppositely charged ions are attracted to each other, thus Cl^- (the symbolic representation of the chloride ion) and Na^+ (the symbol for the sodium ion, using the Greek word *natrium*) form an ionic bond, becoming the molecule sodium chloride. Ionic bonds generally form between elements in Group I (having one electron in their outer shell) and Group VIIa (having seven electrons in their outer shell). Such bonds are relatively weak, and tend to disassociate in water, producing solutions that have both Na and Cl ions.

Top: Formation of a crystal of sodium chloride. Each positively charged sodium ion is surropunded by six negatively charged chloride ions; likewise each negatively charged chloride ion is surrounded by six positively charged sodium ions. The overall effect is electrical neutrality

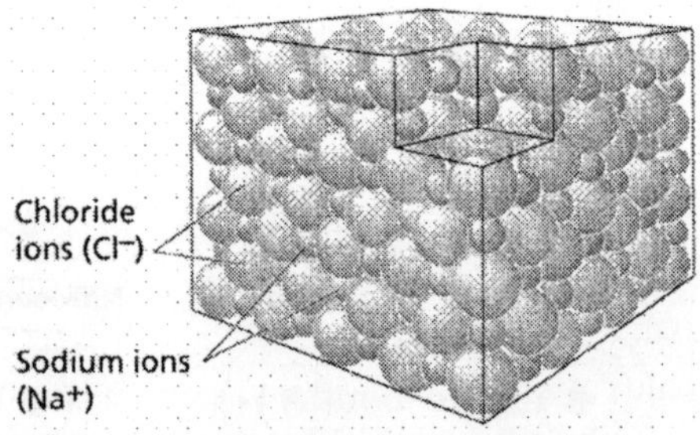

Covalent bonds form when atoms share electrons. Since electrons move very fast they can be shared, effectively filling or emptying the outer shells of the atoms involved in the bond. Such bonds are referred to as electron-sharing bonds. An analogy can be made to child custody: the children are like electrons, and tend to spend some time with one parent and the rest of their time with the other parent. In a covalent bond, the electron clouds surrounding the atomic nuclei overlap.

Formation of a covalent bond between two Hydrogen atoims.

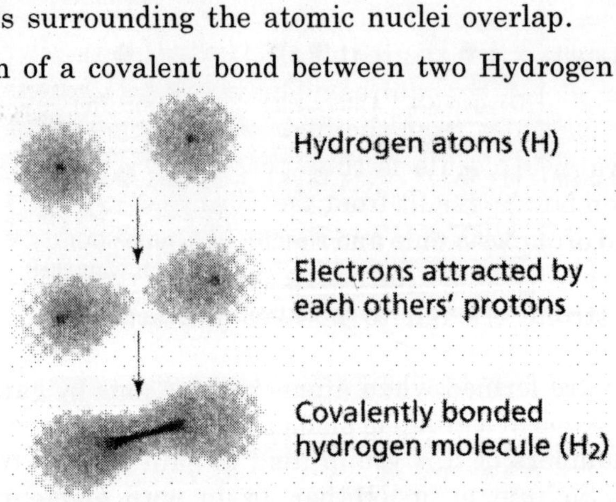

Carbon (C) is in Group IVa, meaning it has four electrons in its outer shell. Thus to become a "happy atom", Carbon can either gain or lose four electrons. By sharing the electrons with other atoms, Carbon can become a happy atom,. alternately filling and emptying its outer shell.

Formation of covalent bonds in methane. Carbon needs to share four electrons, in effect it has four slots. Each hydrogen provides an electron to each of these slots. At the same time each hydrogen needs to fill one slot, which is done by sharing an electron with the carbon

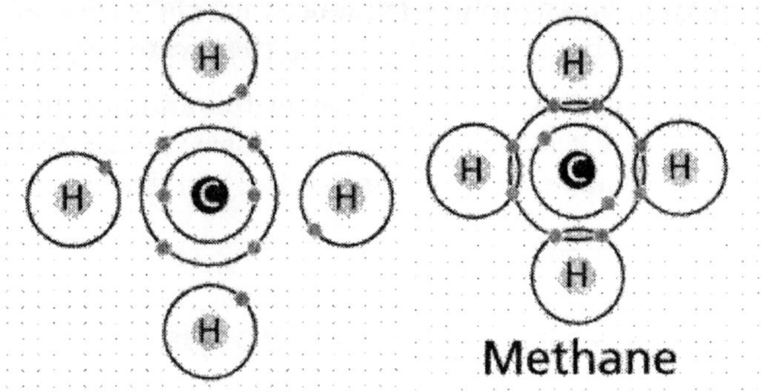

The molecule methane (chemical formula CH_4) has four covalent bonds, one between Carbon and each of the four Hydrogens. Carbon contributes an electron, and Hydrogen contributes an electron. The sharing of a single electron pair is termed a single bond. When two pairs of electrons are shared, a double bond results, as in carbon dioxide. Triple bonds are known, wherein three pairs (six electrons total) are shared as in acetylene gas or nitrogen gas.

Ways of Representing Covalent Bonds

Sometimes electrons tend to spend more time with one atom in the bond than with the other. In such cases a polar covalent bond develops. Water (H_2O) is an example. Since the electrons spend so much time with the oxygen (oxygen having a greater electronegativity, or electron affinity) that end of the molecule acquires a slightly negative charge. Conversely, the loss of the electrons from the hydrogen end leaves a slightly positive charge. The water molecule is thus polar, having positive and negative sides. Hydrogen bonds, result from the weak electrical attraction between the positive end of one molecule and the negative end of another. Individually these bonds are very weak, although taken in a large enough quantity, the result is strong enough to hold molecules together or in a three-dimensional shape.

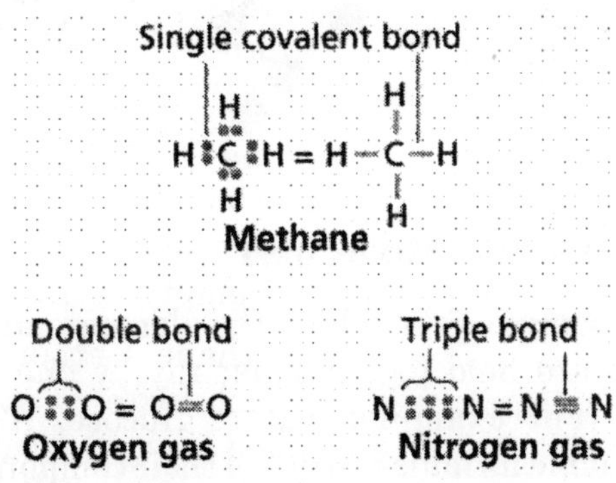

Top: Formation of a hydrogen bond between the hydrogen side of one water molecule and the oxygen side of another water molecule.

Bottom: The presence of polar areas in the amino acids that make-up a protein allows for hydrogen bonds to form, giving the molecule a three-dimensional shape that is often vital to that protein's proper functioning.

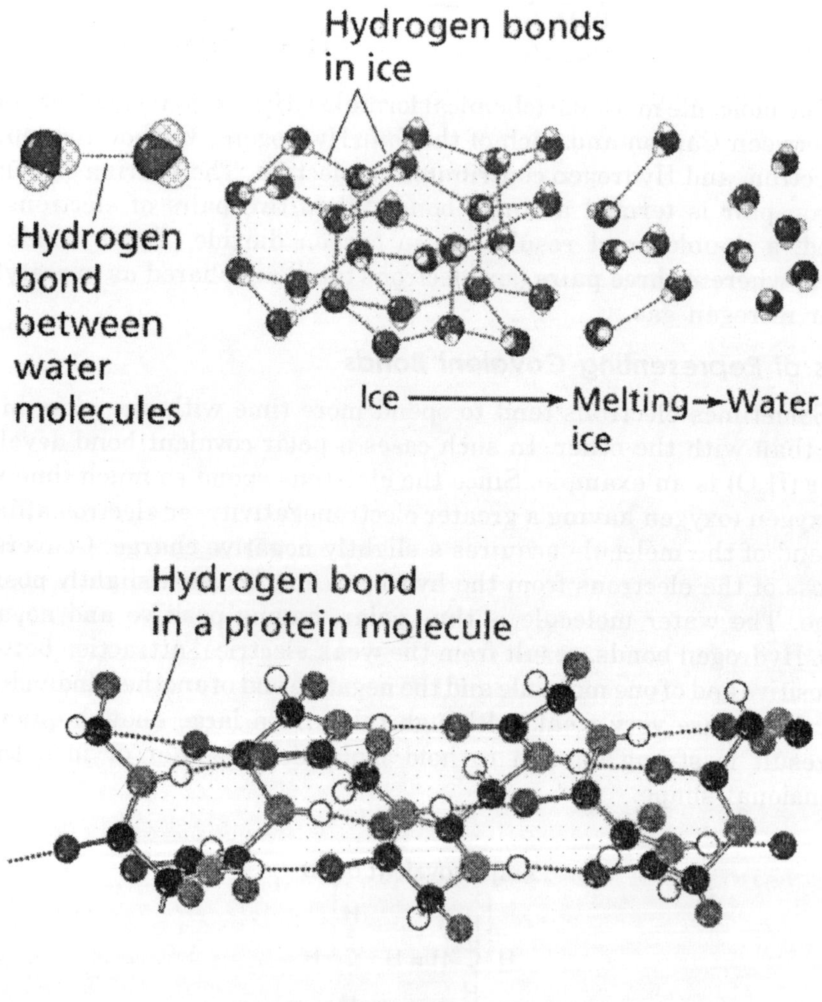

Chemical Reactions and Molecules

Molecules are compounds in which the elements are in definite, fixed ratios. Those atoms are held together usually by one of the three types of chemical bonds. For example: water, glucose, ATP. Mixtures are compounds with variable formulas/ratios of their components. For example:

soil. Molecular formulas are an expression in the simplest whole-number terms of the composition of a substance. For example, the sugar glucose has 6 Carbons, 12 hydrogens, and 6 oxygens per repeating structural unit. The formula is written $C_6H_{12}O_6$.

Determination of molecular weights by addition of the weights of the atoms that make up the molecule.

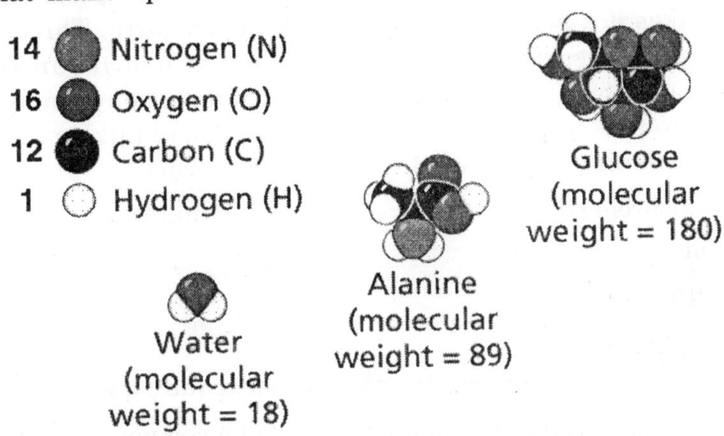

Chemical reactions occur in nature, and some also can be performed in a laboratory setting. One such reaction is diagrammed. Chemical equations are linear representations of how these reactions occur. Combination reactions occur when two separate reactants are bonded together, e.g. A + B ——> AB. Disassociation reactions occur when a compound is broken into two products, e.g. AB ——> A + B.

Diagram of a chemical reaction: the combustion of propane with oxygen, resulting in carbon dioxide, water, and energy (as heat and light). This chemical reaction takes place in a camping stove as well as in certain welding torches

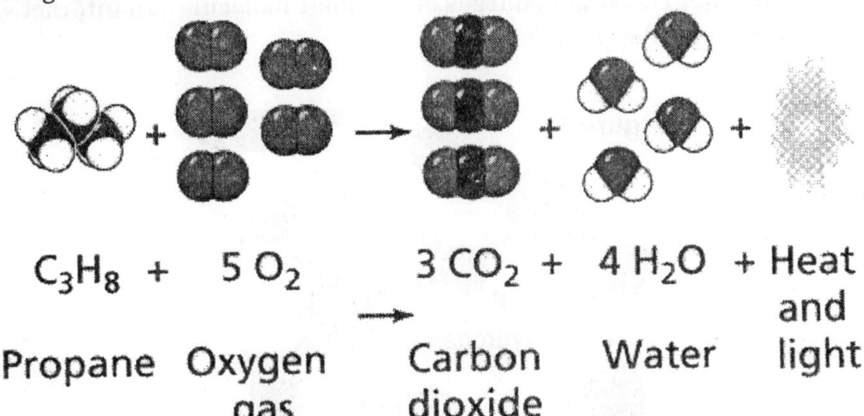

Biological systems, while unique to each species, are based on the

chemical bonding properties of carbon. Major organic chemicals (those associated with or formed by the actions of living things) usually include some ratios of the following elements: C, H, N, O, P, S.

Water—the Fluid of Life

Water is indispensable to life as we know it. The special chemistry of water makes it a major component of all life on earth. Three quarters of the surface of the earth is covered by water and we humans are composed of approximately 66% water. Water has many unusual physical properties that result from its unique chemical structure. Before talking about these properties, it is first necessary to define some terms.

Physical properties of matter include such things as:

- colour
- boiling point
- freezing point
- any physical characteristics that do not require a chemical change.

Chemical properties are characteristics of a compound that relate to its ability to react chemically with another substance. An example of one of the important chemical properties of water is its ability to form hydrogen bonds with other water molecules. Hydrogen bonds are chemical bonds formed between the hydrogen on one molecule and an electronegative element such as oxygen or nitrogen on a different molecule.

Oxygen is an electronegative element and hydrogen is an electropositive element. Since water is composed of one oxygen atom and two hydrogens, water molecules tend to develop slight negative and positive charges at different ends of the molecule. This charge differentiation makes water a polar molecule. The slight charges of the polar molecule can interact with other polar molecules or ions.

Figure 1

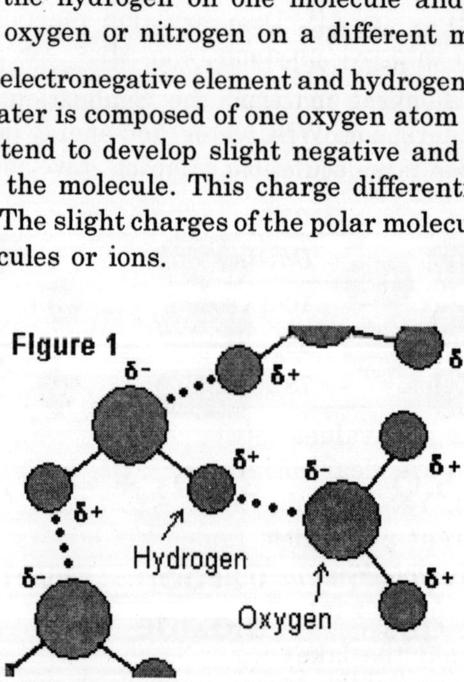

The two hydrogens on one end of the water molecule have partial positive charges (denoted by the symbol d +). The oxygen at the other end of the molecule has a partial negative charge (denoted by the symbol d -). Because of these partial positive and negative charges, water molecules can form hydrogen bonds with other water molecules. One of the partially positive hydrogens will form a weak bond with the partially negative oxygen of another water molecule.

The ability of water molecules to hydrogen bond with other water molecules gives water special physical properties. The hydrogen bonds between water molecules allow water to form a highly ordered lattice structure in the solid state. When water freezes a crystal lattice structure forms. This lattice structure is what gives snowflakes and ice crystals their beautiful design. The polar, ordered nature of water makes it an excellent solvent. It is often referred to as the universal solvent, although not everything is soluble in water obviously!

A solvent is an agent that dissolves, or solubilizes, a substance called a solute. If something is soluble in water it dissolves in it. However, solubilizing a substance does not cause it to vanish. As you certainly know, when sugar dissolves in water the water tastes sweet, indicating the presence of the sugar. When the polar water molecules orient themselves around the partially charged polar or fully charged ionic molecules of the solute, the solute has been solubilized in the water.

The partial positive charges of the hydrogens on the polar water molecules surround each negatively charged chlorine ion, while the partial negative charges of the oxygen on the polar water molecules surround each positively charged sodium ion. When water molecules orient themselves around charged or polar molecules, spheres of hydration, or hydration shells, are formed.

Compound	*Melting Pt.*	*Boiling Pt.*	*Heat of Vaporization*
Water	0° Celsius	100° Celsius	540 Cal/g
Methanol	-98° Celsius	65° Celsius	263 Cal/g
Ethanol	-117° Celsius	78° Celsius	204 Cal/g

Note the wide range in values between these three seemingly similar liquid compounds. All are clear coloured, polar solvents, and all contain oxygen and hydrogen. Yet, the table above shows that these substances have extremely different properties. Imagine if lakes were composed of either methanol or ethanol. What might happen in the heat of the summer? What would happen in the winter?

If you guessed that the lakes would evaporate very quickly in the summer and fail to freeze in the winter you would be correct! As you may imagine, this would have a devastating effect on pond life. The formation of ice is very important to many aquatic organisms as a form of protection against extreme cold and rapid temperature fluctuations.

Water has some additional physical properties that also make it essential to life.

- Cohesion- the molecular attraction between the particles within a substance.
- Adhesion- the molecular attraction between a substance and the surface it contacts.
- Capillarity- a characteristic of water that results from both the cohesive and adhesive properties. Capillarity allows water to rise in a small tube above the level of the surrounding water.
- Surface tension- a characteristic of water that results from the cohesion between water molecules, and allows the surface of water to have some strength to support objects.

pH and Water

In addition to all of its other properties, water can act as either a weak acid or a weak base. How can water act as both? First we must define acids and bases. For our purposes here we will use the simplest definition:

An acid is a substance that can donate hydrogen ions (H+) to the surrounding aqueous environment.

A base is a substance that can absorb or accept hydrogen ions from the surrounding aqueous environment.

An acid will increase the hydrogen ion concentration of an aqueous solution in which it is solubilized. A base will decrease the hydrogen ion concentration of an aqueous solution in which it is solubilized. Acids and bases must be dissolved in aqueous solution in order to dissociate and become active as hydrogen donors or acceptors.

Before we go on we should better define the hydrogen ion (H+). Remember that the element hydrogen is composed of one electron and one proton. Hydrogen normally exists in nature as a gas composed of two hydrogen atoms covalently bonded together. A hydrogen ion is one hydrogen atom that has lost its electron, and therefore is merely a single proton. Knowing what you know about atoms and what makes them stable or unstable, do you suppose that a single proton by itself would be very stable?

Of course it wouldn't! Hydrogen ions can only exist when their positive charges can be neutralized by opposite charges. For example, a hydrogen ion can be present in water when the negative ends of the polar water molecules can surround the positively charged hydrogen ion in a sphere of hydration.

Ionization or separation of water into hydrogen ions (H+) and hydroxide ions (OH-) occurs to a very small extent in water at all times. Water acts as a hydrogen donor (an acid) when a molecule of water separates and releases a hydrogen ion. Water acts as a hydrogen acceptor (a base) when one of the hydroxide ions accepts a hydrogen ion to form a complete molecule of water. The following equation represents the ionization of water: $H_2O \longrightarrow H^+ + OH^-$ Ionization of water causes the neutral water molecules to separate into charged ions.

Acids and bases are classified according to their ability to donate or accept hydrogen ions. A strong acid can donate hydrogen ions easily when dissolved in aqueous solution. A strong base can accept hydrogen ions easily. Now that we have defined acids and bases chemically, let's find out what effects they have on biological systems.

As you already know, charged particles, or ions, are highly reactive. This basically means that ions are looking for other molecules to react with so they may attain a more stable electron configuration. When charged particles react with biomolecules such as protein and nucleic acid, they can have devastating effects upon the organism. Since acids and bases can dissociate easily into charged hydrogen ions and hydroxide ions, they have the potential to be very destructive to living tissues. Hydrochloric acid, a strong acid, can be highly destructive to organic matter when in its concentrated form. The same is true for sodium hydroxide, a strong base. Few people realize that strong bases are just as dangerous as strong acids.

Most oven cleaners (the foamy substances that eat encrusted food on your oven walls) contain concentrated sodium hydroxide base. Many drain openers are often concentrated acids. Drain openers operate by breaking down organic materials like hair that may clog the drain. Knowing what strong acids and bases due to the organic materials in dirty ovens and clogged drains should give you a hint about their potential for tissue destruction. However, even weak and dilute acids and bases can have a devastating effect when not compensated for inside living cells and tissues.

Although we have not yet discussed the role of the major biological molecules I think most of us know that our cells and tissues stay alive by performing many complex biochemical reactions. Biological life revolves around a class of molecules called proteins. Proteins are very sensitive to even a slight increase or decrease in the hydrogen ion concentration (pH) of the surrounding solution. So how can we be exposed to and ingest acidic and basic compounds all the time and yet not increase and decrease the hydrogen ion concentration in our blood, tissues, and cells?

Buffers!

A buffer is a substance that can accept hydrogen ions when there are too many present or release hydrogen ions when there are too few present. In this way buffers maintain the pH at a relatively constant level.

pH is a measure of how many hydrogen ions are present in a substance. pH actually represents the number of hydrogen ions per unit volume of a liquid, or the concentration of hydrogen ions. Here's an example of how concentration works: one package of Kool-Aid makes two quarts. If you place one cup of sugar in that two quarts you have a specific concentration of sugar in your Kool-Aid. If you place two cups of sugar in that two quarts you would have increased the concentration of sugar 2-fold, or 2X. Buffers are essential to all organisms because most biochemical reactions can only occur at neutral pH's of about seven. A substance with a neutral pH has just the right amount of hydrogen ions present for biochemical reactions to occur. Buffers maintain pH by absorbing or releasing hydrogen ions when necessary to maintain a specific hydrogen ion concentration. In our laboratory we perform many extractions of biologically active proteins which we later use in an in vitro reaction. We use buffering agents in all of our liquids so that we can maintain a near neutral pH, similar to the pH found in most biological systems. As a matter of fact, scientists almost universally refer to these liquids used in their laboratories as buffers even though it is technically incorrect because most of them contain components in addition to the buffering compound. Buffers are indispensable in all aqueous solutions in which biochemical reactions are taking place, whether in a living organism or in the laboratory.

Is water very susceptible to a pH change? Yes! Add only one drop of concentrated hydrochloric acid to a liter (approximately a quart) of pure water and the pH will be reduced from seven to approximately one. A pH of seven is neutral, and a pH of one is very acidic. What really is the difference between a solution with a pH of seven and a solution with a pH of one? As we have noted before the pH represents the concentration of hydrogen ions in a specific volume of a substance. pH is defined as the negative log of the hydrogen ion concentration as represented by this equation: -log [H+]. The brackets indicate concentration of H+.

Right now you might be thinking: *negative log of what?!* Relax, it's pretty simple.

The log of a number q in base 10 is equal to the exponent of q when written in exponential notation. For example, the log of 100 is equal to 2, because 100 written in exponential notation is 1×10^2. When figuring pH, we use the negative log. It is the same as taking the log and then multiplying it by negative one.

But wait, pH's are not negative numbers! That is because the pH represents a small number from 0.1 (1×10^{-1}) to 0.00000000000001 (1×10^{-14}). These numbers correspond to a pH of one and a pH of fourteen. As you can see, the log of 1×10^{-14} is -14, but the negative log of 1×10^{-14} is 14. The definition for pH was fabricated to simplify the representation of hydrogen ion concentration. pH values represent a kind of shorthand value for acidic and basic solutions.

Note that the higher the concentration of hydrogen ions, the smaller the exponent, and therefore the smaller the pH value. For this reason, acids will have small pH values, such as a pH of one for a strong acid. Bases will have large pH values because their hydrogen ion concentration is smaller. A solution with a hydrogen ion concentration of 1×10^{-1} has a relatively large number of hydrogen ions. This solution is a strong acid of pH one. A solution with a hydrogen ion concentration of 1×10^{-14} has a very small number of hydrogen ions. This solution is a strong base with pH fourteen.

The Polarity of Water

Water has a simple molecular structure. It is composed of one oxygen atom and two hydrogen atoms. Each hydrogen atom is covalently bonded to the oxygen via a shared pair of electrons. Oxygen also has two unshared pairs of electrons. Thus there are 4 pairs of electrons surrounding the oxygen atom, two pairs involved in covalent bonds with hydrogen, and two unshared pairs on the opposite side of the oxygen atom. Oxygen is an "electronegative" or electron "loving" atom compared with hydrogen.

Water is a "polar" molecule, meaning that there is an uneven distribution of electron density. Water has a partial negative charge ($\delta-$) near the oxygen atom due the unshared pairs of electrons, and partial positive charges ($\delta+$) near the hydrogen atoms.

An electrostatic attraction between the partial positive charge near the hydrogen atoms and the partial negative charge near the oxygen results in the formation of a hydrogen bond as shown in the illustration. The ability of ions and other molecules to dissolve in water is due to polarity. For example, in the illustration below sodium chloride is shown in its crystalline form and dissolved in water.

Many other unique properties of water are due to the hydrogen bonds. For example, ice floats because hydrogen bonds hold water molecules further apart in a solid than in a liquid, where there is one less hydrogen bond per molecule. The unique physical properties, including a high heat of vaporization, strong surface tension, high specific heat, and nearly universal solvent properties of water are also due to hydrogen bonding. The hydrophobic effect, or the exclusion of compounds containing carbon and hydrogen (nonpolar compounds) is another unique property of water caused by the hydrogen bonds. The hydrophobic effect is particularly important in the formation of cell membranes. The best description is to say that water "squeezes" nonpolar molecules together.

Acids and Bases, Ionization of Water

$$H_2O \xrightarrow{\text{Limited}} H^+ + OH^-$$

- Acid release H^+
- Bases accept H^+.

We define the pH of a solution as the negative logarithm of the hydrogen ion concentration.

- at pH 7.0, a solution is neutral
- at lower pH (1-6), a solution is acidic
- at higher pH (8-14), a solution is basic.

The Major Classes of Biological Molecules

Now that we have introduced two groups of biological molecules, let us discuss all four major groups of biological molecules:

- Proteins
- Nucleic Acids
- Carbohydrates
- Lipids.

These four types of biological molecules compose all life on earth.

Proteins have an incredible variety of functions in living organisms. The functions of proteins include but are not limited to:

- acting as biological catalysts
- forming structural parts of organisms
- participating in cell signal and recognition factors
- acting as molecules of immunity.

Nucleic Acids consist of two distinct, but closely related chemical forms: deoxyribonucleic acid (DNA) and ribonucleic acid (RNA). The main functions of these biomolecules include the storage of all heritable information of all organisms on earth and the conversion of this information into proteins.

Carbohydrates are the major source of food and the major form of energy for most living organisms. When carbohydrates combine together to form polymers they can function as long term food storage molecules, as protective coverings for cells and organisms, as the main structural support for land plants and constituents of many cells and their contents.

Lipids are the major constituents of all membranes in all cells. They also serve as food storage molecules. This class of biological molecules includes the hydrophobic fats, oils and waxes.

Often times students have difficulty remembering the four major classes of biological molecules, so I have a way to help. To recall the four major classes of biological molecules I want you to think of lunch at a hamburger joint. No endorsements intended! Take a hamburger, fries, and a milkshake and you have all the components of the four major classes of biological molecules.

Of all four classes which do you think is most abundant in this meal?

If you guessed LIPIDS you would be right! The hamburger is full of fat, and processors add fat to the hamburger bun. Mayonnaise is oil and egg yolk, which is mostly fat. Milk contains fat (if your shake has any milk product in it), and fat is added to nondairy shakes. As for the fries, we all know why they are called fries!

Okay- we know there are plenty of lipids, but where do the rest of the biological molecules come in?

The hamburger bun is made from wheat flour, which is full of starch, a carbohydrate. The sugar in your shake is the disaccharide sucrose, which is also a carbohydrate. As a matter of fact, the lettuce, onion and tomato on that burger all contain carbohydrates.

Meat is composed of muscle tissue, and muscle tissue is a structural form of protein. Also most of the formerly living things that we eat have some form of protein in them: wheat, milk, potatoes, and so on. All cells contain DNA and RNA, except mature mammalian red blood cells, so DNA and RNA can be found in all organic matter. The lettuce, onion, and tomato contain nucleic acid, and the hamburger meat contains cow DNA.

The Magic of Carbon

Carbon is the chemical element with symbol C and atomic number 6. As a member of group 14 on the periodic table, it is nonmetallic and tetravalent—making four electrons available to form covalent chemical bonds. There are three naturally occurring isotopes, with ^{12}C and ^{13}C being stable, while ^{14}C is radioactive, decaying with a half-life of about 5730 years. Carbon is one of the few elements known since antiquity. The name "carbon" comes from Latin language *carbo*, coal.

There are several allotropes of carbon of which the best known are graphite, diamond, and amorphous carbon. The physical properties of carbon vary widely with the allotropic form. For example, diamond is highly transparent, while graphite is opaque and black. Diamond is among the hardest materials known, while graphite is soft enough to form a streak on paper (hence its name, from the Greek word "to write"). Diamond has a very low electrical conductivity, while graphite is a very good conductor. Under normal conditions, diamond has the highest thermal conductivity of all known materials. All the allotropic forms are solids under normal conditions but graphite is the most thermodynamically stable.

All forms of carbon are highly stable, requiring high temperature to react even with oxygen. The most common oxidation state of carbon in inorganic compounds is +4, while +2 is found in carbon monoxide and other transition metal carbonyl complexes. The largest sources of inorganic carbon are limestones, dolomites and carbon dioxide, but significant quantities occur in organic deposits of coal, peat, oil and methane clathrates. Carbon forms more compounds than any other element, with almost ten million pure organic compounds described to date, which in turn are a tiny fraction of such compounds that are theoretically possible under standard conditions.

Carbon is the 15th most abundant element in the Earth's crust, and the fourth most abundant element in the universe by mass after hydrogen, helium, and oxygen. It is present in all known lifeforms, and in the human body carbon is the second most abundant element by mass (about 18.5%) after oxygen. This abundance, together with the unique diversity of organic compounds and their unusual polymer-forming ability at the temperatures commonly encountered on Earth, make this element the chemical basis of all known life.

Carbon has the ability to form very long chains of interconnecting C-C bonds. This property is called catenation. Carbon-carbon bonds are strong, and stable. This property allows carbon to form an almost infinite number of compounds; in fact, there are more known carbon-containing compounds than all the compounds of the other chemical elements combined except those of hydrogen (because almost all organic compounds contain hydrogen too).

The simplest form of an organic molecule is the hydrocarbon—a large family of organic molecules that are composed of hydrogen atoms bonded to a chain of carbon atoms. Chain length, side chains and functional groups all affect the properties of organic molecules. By IUPAC's definition, all the other organic compounds are functionalized compounds of hydrocarbons.

Carbon occurs in all known organic life and is the basis of organic chemistry. When united with hydrogen, it forms various hydrocarbons which are important to industry as refrigerants, lubricants, solvents, as chemical feedstock for the manufacture of plastics and petrochemicals and as fossil fuels.

When combined with oxygen and hydrogen, carbon can form many groups of important biological compounds including sugars, lignans, chitins, alcohols, fats, and aromatic esters, carotenoids and terpenes. With nitrogen it forms alkaloids, and with the addition of sulphur also it forms antibiotics, amino acids, and rubber products. With the addition of phosphorus to these other elements, it forms DNA and RNA, the chemical-code carriers of life, and adenosine triphosphate (ATP), the most important energy-transfer molecule in all living cells.

Inorganic Compounds

Commonly carbon-containing compounds which are associated with minerals or which do not contain hydrogen or fluorine, are treated separately from classical organic compounds; however the definition is not rigid. Among these are the simple oxides of carbon. The most prominent oxide is carbon dioxide (CO_2). This was once the principal constituent of the paleoatmosphere, but is a minor component of the Earth's atmosphere today. Dissolved in water, it forms carbonic acid (H_2CO_3), but as most compounds with multiple single-bonded oxygens on a single carbon it is unstable. Through this intermediate, though, resonance-stabilized carbonate ions are produced. Some important minerals are carbonates, notably calcite. Carbon disulfide (CS_2) is similar.

The other common oxide is carbon monoxide (CO). It is formed by incomplete combustion, and is a colorless, odorless gas. The molecules each contain a triple bond and are fairly polar, resulting in a tendency to bind permanently to hemoglobin molecules, displacing oxygen, which has a lower binding affinity. Cyanide (CN^-), has a similar structure, but behaves much like a halide ion (pseudohalogen). For example it can form the nitride cyanogen molecule ($(CN)_2$), similar to diatomic halides. Other uncommon oxides are carbon suboxide (C_3O_2), the unstable dicarbon monoxide (C_2O), carbon trioxide (CO_3), cyclopentanepentone (C_5O_5), cyclohexanehexone (C_6O_6), and mellitic anhydride ($C_{12}O_9$).

Carbon Cycle

The carbon cycle is the biogeochemical cycle by which carbon is exchanged among the biosphere, pedosphere, geosphere, hydrosphere, and atmosphere of the Earth. It is one of the most important cycles of the earth and allows for carbon to be recycled and reused throughout the biosphere and all of its organisms.

The carbon cycle was initially discovered by Joseph Priestley and

Antoine Lavoisier, and popularized by Humphry Davy. It is now usually thought of as including the following major reservoirs of carbon interconnected by pathways of exchange:

- The atmosphere
- The terrestrial biosphere, which is usually defined to include fresh water systems and non-living organic material, such as soil carbon.
- The oceans, including dissolved inorganic carbon and living and non-living marine biota,
- The sediments including fossil fuels.
- The Earth's interior, carbon from the Earth's mantle and crust is released to the atmosphere and hydrosphere by volcanoes and geothermal systems.

Carbon forms the backbone of biology for all life on Earth. Complex molecules are made up of carbon bonded with other elements, especially oxygen, hydrogen and nitrogen, and carbon is able to bond with all of these because of its four valence electrons. It is often assumed in astrobiology that if life exists somewhere else in the universe, it will also be carbon based. This assumption is referred to by critics as *carbon chauvinism.*

Characteristics of Carbon as a Basis for Life

The two most important characteristics of carbon as a basis for the chemistry of life, are that it has four valence bonds and that the energy required to make or break a bond is just at an appropriate level for building molecules which are not only stable, but also reactive. The fact that carbon atoms bond readily to other carbon atoms allows for the building of arbitrarily long and complex molecules. There are not many other elements which appear to be even promising candidates for supporting life-like behaviour, but the most frequent alternative suggested is silicon. This is in the same group in the Periodic Table of elements and therefore also has four valence bonds. It also bonds to itself, but generally in the form of crystal lattices, less amenable to a complete source of life, rather than long chains. Its compounds are generally highly stable and do not support the ability readily to re-combine in different permutations in a manner that would plausibly support life-like processes. This speculation of a life based on the chemistry of silicon is clearly distinct from "silicon-based life" in the above sense of artificial intelligence based on electronic processes utilizing silicon integrated circuits.

Key Carbon-based Molecules in the Life Processes

The most notable groups of chemicals used in the processes of living organisms include:

- Proteins, which are the building blocks from which the structures of living organisms are constructed (this includes almost all enzymes, which catalyse organic chemical reactions)
- Nucleic acids, which carry genetic information
- Carbohydrates, which store energy in a form that can be used by living cells
- Fats, which also store energy, but in a more concentrated form, and which may be stored for extended periods in the bodies of animals.

Carbon's importance to plants and animals is extremely important. Its presence within the atmosphere as well as within living matter makes it fundamental to their survival.

Plants

Carbon is essential to plants in order for them to perform photosynthesis—the chemical reaction that plants use to create energy. Plants take carbon from carbon dioxide in the atmosphere and produce food for themselves, as well as water and oxygen for the environment.

Animals

Animals are dependent upon water to drink as well as oxygen in the air in order to breathe. Without carbon, plants would not be able to produce the necessary water and oxygen for animal environments, and would not survive to be a food source for animals.

Organic Compounds

Carbon atoms form molecular bonds at very basic levels. For this reason, important organic compounds such as simple proteins and sugars are easily synthesized in the environment, providing necessary nutrients for plants and animals.

Cell Structure

Due to their physical structure and chemical properties, carbon atoms may form a number of structures and configurations that are key to cell structure and functioning.

The Basis of Life

Carbon atoms are found not only in all plants and animals, but in all other living life forms as well. Without carbon, life as we know it would simply not exist.

3

Basic Chemistry of Nucleic Acid

Living organisms are complex systems. Hundreds of thousands of proteins exist inside each one of us to help carry out our daily functions. These proteins are produced locally, assembled piece-by-piece to exact specifications. An enormous amount of information is required to manage this complex system correctly. This information, detailing the specific structure of the proteins inside of our bodies, is stored in a set of molecules called nucleic acids.

Thermal Properties of DNA

As cells divide it is a necessity that the DNA be copied (replicated), in such a way that each daughter cell acquires the same amount of genetic material. In order for this process to proceed the two strands of the helix must first be separated, in a process termed denaturation. This process can also be carried out in vitro. If a solution of DNA is subjected to high temperature, the H-bonds between bases become unstable and the strands of the helix separate in a process of thermal denaturation.

The base composition of DNA varies widely from molecule to molecule and even within different regions of the same molecule. Regions of the duplex that have predominantly A-T base-pairs will be less thermally stable than those rich in G-C base-pairs. In the process of thermal denaturation, a point is reached at which 50% of the DNA molecule exists as single strands. This point is the melting temperature (T_m), and is characteristic of the base composition of that DNA molecule. The T_m depends upon several factors in addition to the base composition. These include the chemical nature of the solvent and the identities and concentrations of ions in the solution.

When thermally melted DNA is cooled, the complementary strands will again re-form the correct base pairs, in a process is termed annealing or hybridization. The rate of annealing is dependent upon the nucleotide sequence of the two strands of DNA.

Analysis of DNA Structure

Chromatography: Several of the chromatographic techniques available for the characterization of proteins can also be applied to the characterization of DNA. The most commonly used technique is HPLC (high performance liquid chromatography). Affinity chromatographic techniques also can be employed. One common affinity matrix is hydroxyapatite (a form of calcium phosphate), which binds double-stranded DNA with a higher affinity than single-stranded DNA.

Electrophoresis: This procedure can serve the same function with regard to DNA molecules as it does for the analysis of proteins. However, since DNA molecules have much higher molecular weights than proteins, the molecular sieve used in electrophoresis of DNA must be different as well. The material of choice is agarose, a carbohydrate polymer purified from a salt water algae. It is a copolymer of mannose and galactose that when melted and re-cooled forms a gel with pores sizes dependent upon the concentration of agarose. The phosphate backbone of DNA is highly negatively charged, therefore DNA will migrate in an electric field. The size of DNA fragments can then be determined by comparing their migration in the gel to known size standards. Extremely large molecules of DNA (in the range of 30kb–10Mb) are effectively separated in agarose gels using pulsed-field gel electrophoresis (PFGE). This technique employs two or more electrodes, placed orthogonally with respect to the gel, that receive short alternating pulses of current. PFGE allows whole chromosomes and large portions of chromosomes to be analysed.

Carbohydrates

In many ways, our bodies can be thought of as chemical processing plants. Chemicals are taken in, processed through various types of reactions, and then distributed throughout the body to be used immediately or stored for later use. The chemicals used by the body can be divided into two broad categories: macronutrients, those substances that we need to eat regularly in fairly large quantities, and micronutrients, those substances that we need only in small amounts. Three major classes of macronutrients are essential to living organisms: carbohydrates, fats, and proteins.

A carbohydrate is an organic compound which has the empirical formula $C_m(H_2O)_n$; that is, consists only of carbon, hydrogen and oxygen, with a hydrogen:oxygen atom ratio of 2:1 (as in water). Carbohydrates can be viewed as hydrates of carbon, hence their name. Structurally however, it is more accurate to view them as polyhydroxy aldehydes and ketones.

The term is most common in biochemistry, where it is a synonym of saccharide. The carbohydrates (saccharides) are divided into four chemical groupings: monosaccharides, disaccharides, oligosaccharides, and polysaccharides. In general, the monosaccharides and disaccharides, which are smaller (lower molecular weight) carbohydrates, are commonly referred to as sugars. While the scientific nomenclature of carbohydrates is complex, the names of the monosaccharides and disaccharides very often end in the suffix-ose. For example, blood sugar is the monosaccharide glucose, table sugar is the disaccharide sucrose, and milk sugar is the disaccharide lactose.

Carbohydrates perform numerous roles in living things. Polysaccharides serve for the storage of energy (e.g., starch and glycogen) and as structural components (e.g., cellulose in plants and chitin in arthropods). The 5-carbon monosaccharide ribose is an important component of coenzymes (e.g., ATP, FAD, and NAD) and the backbone of the genetic molecule known as RNA. The related deoxyribose is a component of DNA. Saccharides and their derivatives include many other important biomolecules that play key roles in the immune system, fertilization, preventing pathogenesis, blood clotting, and development.

In food science and in many informal contexts, the term carbohydrate often means any food that is particularly rich in the complex carbohydrate starch (such as cereals, bread and pasta) or simple carbohydrates, such as sugar (found in candy, jams and desserts).

Structure

Formerly the name "carbohydrate" was used in chemistry for any compound with the formula $C_m(H_2O)_n$. Following this definition, some chemists considered formaldehyde CH_2O to be the simplest carbohydrate, while others claimed that title for glycolaldehyde Today the term is generally understood in the biochemistry sense, which excludes compounds with only one or two carbons.

Natural saccharides are generally built of simple carbohydrates called monosaccharides with general formula $(CH_2O)_n$ where *n* is three or more. A typical monosaccharide has the structure H-$(CHOH)_x$(C=O)-$(CHOH)_y$-H, that is, an aldehyde or ketone with many hydroxyl groups added, usually one on each carbon atom that is not part of the aldehyde or ketone functional group. Examples of monosaccharides are glucose, fructose, and glyceraldehyde. However, some biological substances commonly called "monosaccharides" do not conform to this formula (e.g., uronic acids and deoxy-sugars such as fucose), and there are many chemicals that do conform to this formula but are not considered to be monosaccharides (e.g., formaldehyde CH_2O and inositol $(CH_2O)_6$). The open-chain form of a monosaccharide often coexists with a closed ring form where the aldehyde/ ketone carbonyl group carbon (C=O) and hydroxyl group (-OH) react forming a hemiacetal with a new C-O-C bridge.

Monosaccharides can be linked together into what are called polysaccharides (or oligosaccharides) in a large variety of ways. Many carbohydrates contain one or more modified monosaccharide units that have had one or more groups replaced or removed. For example, deoxyribose, a component of DNA, is a modified version of ribose; chitin is composed of repeating units of N-acetylglucosamine, a nitrogen-containing form of glucose.

Monosaccharides

Monosaccharides are the most basic units of biologically important carbohydrates. They are the simplest form of sugar and are usually colorless, water-soluble, crystalline solids. Some monosaccharides have a sweet taste. Examples of monosaccharides include glucose (dextrose), fructose (levulose), galactose, xylose and ribose. Monosaccharides are the building blocks of disaccharides such as sucrose and polysaccharides (such as cellulose and starch). Further, each carbon atom that supports a hydroxyl group (except for the first and last) is chiral, giving rise to a number of isomeric forms all with the same chemical formula. For instance, galactose and glucose are both aldohexoses, but have different chemical and physical properties.

Structure

With few exceptions (e.g., deoxyribose), monosaccharides have the chemical formula $C_x(H_2O)_y$ with the chemical structure $H(CHOH)_nC{=}O(CHOH)_mH$. If n or m is zero, it is an aldehyde and is termed an aldose; otherwise, it is a ketone and is termed a ketose. Monosaccharides contain either a ketone or aldehyde functional group, and hydroxyl groups on most or all of the non-carbonyl carbon atoms.

Monosaccharide Momenclature

Monosaccharides are classified by the number of carbon atoms they contain:

- Triose, 3 carbon atoms
- Tetrose, 4 carbon atoms
- Pentose, 5 carbon atoms
- Hexose, 6 carbon atoms
- Heptose, 7 carbon atoms.

Some monosaccharides exist in equilibrium between its open chain and the cyclic hemiacetal forms:

Aldoses can form:

- Pyranose results from the intramolecular nucleophilic addition of the C5-OH group to the C1 carbonyl group, yielding a 6-membered ring.

Ketoses can form:

- Pyranose results from the intramolecular nucleophilic addition of the -OH at C6 to the C1 carbonyl group, forming a 6-membered ring.
- Furanose results from the intramolecular nucleophilic addition of the C4-OH group at C5 and to the C1 carbonyl group, yielding a 5-membered ring.

Monosaccharides are classified by the type of carbonyl group they contain:

- Aldose, -CHO (aldehyde)
- Ketose, C=O (ketone).

Isomerism

The total number of possible stereoisomers of one compound (n) is dependent on the number of stereogenic centres (c) in the molecule. The upper limit for the number of possible stereoisomers is $n = 2^c$. The only monosaccharide without an isomer is dihydroxyacetone or DHA.

Monosaccharides are classified according to their molecular configuration at the chiral carbon furthest removed from the aldehyde or ketone group. The chirality at this carbon is compared to the chirality of carbon 2 on glyceraldehyde. If it is equivalent to D-glyceraldehyde' s C2, the sugar is D; if it is equivalent to L-glyceraldehyde's C2, the sugar is L. Due to the chirality of the sugar molecules, an aqueous solution of a D or L saccharides will rotate light.

D-glyceraldehyde causes polarized light to rotate clockwise (dextrorotary); L-glyceraldehyde causes polarized light to rotate counterclockwise (levorotary). Unlike glyceraldehyde, D/L designation on more complex sugars is not associated with their direction of light rotation. Since more complex sugars contain multiple chiral carbons, the direction of light rotation cannot be predicted by the chirality of the carbon that defines D/L nomenclature.

- D, configuration as in D-glyceraldehyde
- L, configuration as in L-glyceraldehyde.

All these classifications can be combined, resulting in names like *D-aldohexose* or *ketotriose.*

Derivatives

A large number of biologically important modified monosaccharides exist:

- Amino sugars such as:
  - o Galactosamine
  - o Glucosamine
  - o Sialic acid
  - o *N*-Acetylglucosamine
- Sulfosugars such as:
  - o Sulfoquinovose.

Monosaccharides are the simplest carbohydrates in that they cannot be hydrolysed to smaller carbohydrates. They are aldehydes or ketones with two or more hydroxyl groups. The general chemical formula of an unmodified monosaccharide is $(C\cdot H_2O)_n$, literally a "carbon hydrate." Monosaccharides are important fuel molecules as well as building blocks for nucleic acids. The smallest monosaccharides, for which n = 3, are dihydroxyacetone and D- and L-glyceraldehyde.

Classification of monosaccharides

Monosaccharides are classified according to three different characteristics: the placement of its carbonyl group, the number of carbon atoms it contains, and its chiral handedness. If the carbonyl group is an aldehyde, the monosaccharide is an aldose; if the carbonyl group is a ketone, the monosaccharide is a ketose. Monosaccharides with three carbon atoms are called trioses, those with four are called tetroses, five are called pentoses, six are hexoses, and so on. These two systems of classification are often combined. For example, glucose is an aldohexose (a six-carbon aldehyde), ribose is an aldopentose (a five-carbon aldehyde), and fructose is a ketohexose (a six-carbon ketone).

Each carbon atom bearing a hydroxyl group (-OH), with the exception of the first and last carbons, are asymmetric, making them stereocenters with two possible configurations each (R or S). Because of this asymmetry, a number of isomers may exist for any given monosaccharide formula. The aldohexose D-glucose, for example, has the formula $(C\cdot H_2O)_6$, of which all but two of its six carbons atoms are stereogenic, making D-glucose one of $2^4 = 16$ possible stereoisomers. In the case of glyceraldehyde, an aldotriose, there is one pair of possible stereoisomers, which are enantiomers and epimers. 1,3-dihydroxyacetone, the ketose corresponding to the aldose glyceraldehyde, is a symmetric molecule with no stereocenters). The assignment of D or L is made according to the orientation of the asymmetric carbon furthest from the carbonyl group: in a standard Fischer projection if the hydroxyl group is on the right the molecule is a D sugar, otherwise it is an L sugar. The "D-" and "L-" prefixes should not be confused with "d-" or "l-", which indicate the direction that the sugar rotates plane polarized light. This usage of "d-" and "l-" is no longer followed in carbohydrate chemistry.

Ring-straight chain isomerism

The aldehyde or ketone group of a straight-chain monosaccharide will react reversibly with a hydroxyl group on a different carbon atom to form a hemiacetal or hemiketal, forming a heterocyclic ring with an oxygen bridge between two carbon atoms. Rings with five and six atoms are called furanose and pyranose forms, respectively, and exist in equilibrium with the straight-chain form.

During the conversion from straight-chain form to cyclic form, the carbon atom containing the carbonyl oxygen, called the anomeric carbon, becomes a stereogenic centre with two possible configurations: The oxygen atom may take a position either above or below the plane of the ring. The resulting possible pair of stereoisomers are called anomers. In the *α anomer*, the -OH substituent on the anomeric carbon rests on the opposite side (trans) of the ring from the CH_2OH side branch. The alternative form, in which the CH_2OH substituent and the anomeric hydroxyl are on the same side (cis) of the plane of the ring, is called the *β anomer*. You can remember that the β anomer is cis by the mnemonic, "It's always better to βe up". Because the ring and straight-chain forms readily interconvert, both anomers exist in equilibrium. In a Fischer Projection, the α anomer is represented with the anomeric hydroxyl group *trans* to the CH_2OH and *cis* in the β anomer.

Use in living organisms

Monosaccharides are the major source of fuel for metabolism, being used both as an energy source (glucose being the most important in nature) and in biosynthesis. When monosaccharides are not immediately needed by many cells they are often converted to more space efficient forms, often polysaccharides. In many animals, including humans, this storage form is glycogen, especially in liver and muscle cells. In plants, starch is used for the same purpose.

Disaccharides

Two joined monosaccharides are called a disaccharide and these are the simplest polysaccharides. Examples include sucrose and lactose. They are composed of two monosaccharide units bound together by a covalent bond known as a glycosidic linkage formed via a dehydration reaction, resulting in the loss of a hydrogen atom from one monosaccharide and a hydroxyl group from thc other. The formula of unmodified disaccharides is $C_{12}H_{22}O_{11}$. Although there are numerous kinds of disaccharides, a handful of disaccharides are particularly notable.

Sucrose, pictured to the right, is the most abundant disaccharide, and the main form in which carbohydrates are transported in plants. It is composed of one D-glucose molecule and one D-fructose molecule. The systematic name for sucrose, *O*-α-D-glucopyranosyl-(1→2)-D-fructofuranoside, indicates four things:

- Its monosaccharides: glucose and fructose
- Their ring types: glucose is a pyranose, and fructose is a furanose
- How they are linked together: the oxygen on carbon number 1 (C1) of α-D-glucose is linked to the C2 of D-fructose.
- The *-oside* suffix indicates that the anomeric carbon of both monosaccharides participates in the glycosidic bond.

Lactose, a disaccharide composed of one D-galactose molecule and one D-glucose molecule, occurs naturally in mammalian milk. The systematic name for lactose is *O*-β-D-galactopyranosyl-(1→4)-D-glucopyranose. Other notable disaccharides include maltose (two D-glucoses linked α-1,4) and cellulobiose (two D-glucoses linked β-1,4). disaccharides can be classified into two types they are reducing and non-reducing disaccahrides if the functional group is present in bonding with another sugar unit it is called as reducing disaccharide.

Oligosaccharides and polysaccharides

Oligosaccharides and polysaccharides are composed of longer chains of monosaccharide units bound together by glycosidic bonds. The distinction between the two is based upon the number of monosaccharide units present in the chain. Oligosaccharides typically contain between three and ten monosaccharide units, and polysaccharides contain greater than ten monosaccharide units. Definitions of how large a carbohydrate must be to fall into each category vary according to personal opinion. Examples of oligosaccharides include the disaccharides mentioned above, the trisaccharide raffinose and the tetrasaccharide stachyose.

Oligosaccharides are found as a common form of protein posttranslational modification. Such posttranslational modifications include the Lewis and ABO oligosaccharides responsible for blood group classifications and so of tissue incompatibilities, the alpha-Gal epitope responsible for hyperacute rejection in xenotransplantation, and O-GlcNAc modifications.

Polysaccharides represent an important class of biological polymers. Their function in living organisms is usually either structure- or storage-related. Starch (a polymer of glucose) is used as a storage polysaccharide in plants, being found in the form of both amylose and the branched amylopectin. In animals, the structurally similar glucose polymer is the more densely branched glycogen, sometimes called 'animal starch'. Glycogen's properties allow it to be metabolized more quickly, which suits the active lives of moving animals.

Cellulose and chitin are examples of structural polysaccharides. Cellulose is used in the cell walls of plants and other organisms, and is claimed to be the most abundant organic molecule on earth. It has many uses such as a significant role in the paper and textile industries, and is used as a feedstock for the production of rayon (via the viscose process), cellulose acetate, celluloid, and nitrocellulose. Chitin has a similar structure, but has nitrogen-containing side branches, increasing its strength. It is found in arthropod exoskeletons and in the cell walls of some fungi. It also has multiple uses, including surgical threads. Other polysaccharides include callose or laminarin, chrysolaminarin, xylan, arabinoxylan, mannan, fucoidan and galactomannan. Carbohydrates are the main energy source for the human body. Chemically, carbohydrates are organic molecules in which carbon, hydrogen, and oxygen bond together in the ratio: $C_x(H_2O)_y$, where x and y are whole numbers that differ depending on the specific carbohydrate to which we are referring. Animals (including humans) break down carbohydrates during the process of metabolism to release energy. For example, the chemical metabolism of the sugar glucose is shown below:

$$C_6H_{12}O_6 + 6\ O_2\ 6\ CO_2 + 6\ H_2O + \text{energy}$$

Animals obtain carbohydrates by eating foods that contain them, for example potatoes, rice, breads, and so on. These carbohydrates are manufactured by plants during the process of photosynthesis. Plants harvest energy from sunlight to run the reaction just described in reverse:

$$6\ CO_2 + 6\ H_2O + \text{energy (from sunlight)}\ C_6H_{12}O_6 + 6\ O_2$$

A potato, for example, is primarily a chemical storage system containing glucose molecules manufactured during photosynthesis. In a potato, however, those glucose molecules are bound together in a long chain. As it turns out, there are two types of carbohydrates, the simple sugars and those carbohydrates that are made of long chains of sugars- the complex carbohydrates.

Simple Sugars

All carbohydrates are made up of units of sugar (also called saccharide units). Carbohydrates that contain only one sugar unit (monosaccharides) or two sugar units (disaccharides) are referred to as simple sugars. Simple sugars are sweet in taste and are broken down quickly in the body to release energy. Two of the most common monosaccharides are glucose and fructose. Glucose is the primary form of sugar stored in the human body for energy. Fructose is the main sugar found in most fruits. Both glucose and fructose have the same chemical formula ($C_6H_{12}O_6$); however, they have different structures, as shown (note: the carbon atoms that sit in the "corners" of the rings are not labelled):

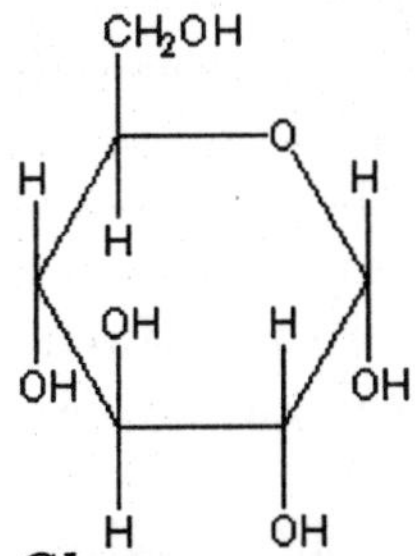

Glucose

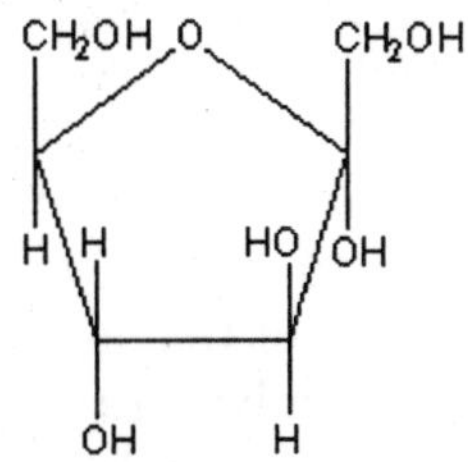

Fructose

Disaccharides have two sugar units bonded together. For example, common table sugar is sucrose, a disaccharide that consists of a glucose unit bonded to a fructose unit:

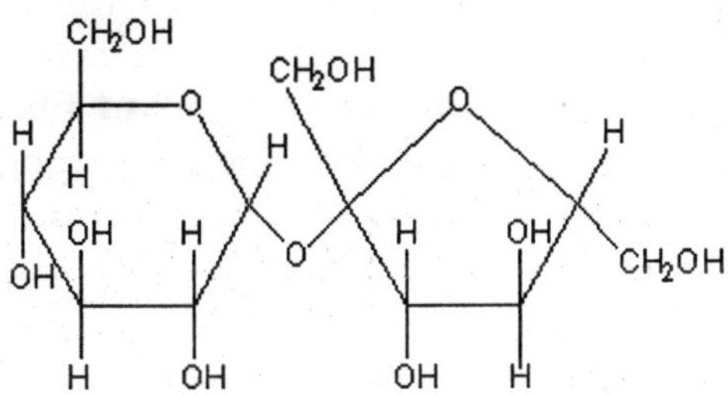

Sucrose

Complex carbohydrates Complex carbohydrates are polymers of the simple sugars. In other words, the complex carbohydrates are long chains of simple sugar units bonded together (for this reason the complex carbohydrates are often referred to as polysaccharides). The potato we discussed earlier actually contains the complex carbohydrate starch. Starch is a polymer of the monosaccharide glucose:

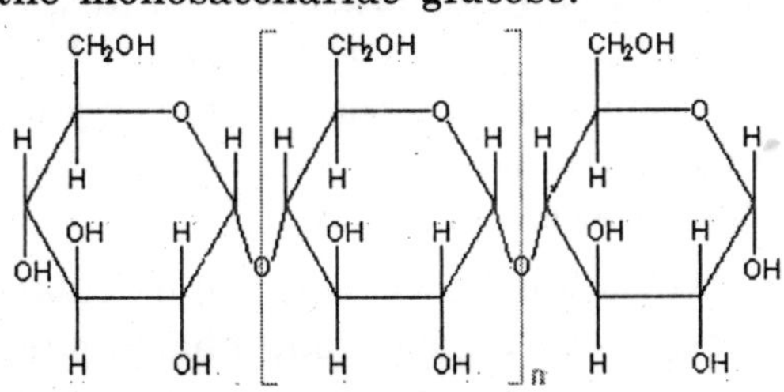

Starch

(n is the number of repeating glucose units and ranges in the 1,000's)

Starch is the principal polysaccharide used by plants to store glucose for later use as energy. Plants often store starch in seeds or other specialized organs; for example, common sources of starch include rice, beans, wheat, corn, potatoes, and so on. When humans eat starch, an enzyme that occurs in saliva and in the intestines called amylase breaks the bonds between the repeating glucose units, thus allowing the sugar to be absorbed into the bloodstream. Once absorbed into the bloodstream, the human body distributes glucose to the areas where it is needed for energy or stores it as its own special polymer-glycogen. Glycogen, another polymer of glucose, is the polysaccharide used by animals to store energy. Excess glucose is bonded together to form glycogen molecules, which the animal stores in the liver and muscle tissue as an "instant" source of energy. Both starch and glycogen are polymers of glucose; however, starch is a long, straight chain of glucose units, whereas glycogen is a branched chain of glucose units.

Another important polysaccharide is cellulose. Cellulose is yet a third polymer of the monosaccharide glucose. Cellulose differs from starch and glycogen because the glucose units form a two-dimensional structure, with hydrogen bonds holding together nearby polymers, thus giving the molecule added stability. Cellulose, also known as plant fiber, cannot be digested by human beings, therefore cellulose passes through the digestive tract without being absorbed into the body. Some animals, such as cows and termites, contain bacteria in their digestive tract that help them to digest cellulose. Cellulose is a relatively stiff material, and in plants it is used as a structural molecule to add support to the leaves, stem, and other plant parts. Despite the fact that it cannot be used as an energy source in most animals, cellulose fiber is essential in the diet because it helps exercise the digestive track and keep it clean and healthy.

How Organisms Convert Food to Energy

Do you ever wonder how organisms convert food to energy?

We all know that organisms must take in nutrients to survive. This self-instructional unit describes two of the most important processes in food to energy conversion. These processes, called glycolysis and respiration, use the energy of the glucose molecule to manufacture ATP, the most widely used energy molecule of the cell.

ATP is an abbreviation for the nucleotide Adenosine triphosphate. This nucleotide consists of a ribose, a phosphate and the nitrogenous base adenine. This is the same nucleotide used to manufacture RNA. Don't be surprised by the fact that ATP is very energetic. All of the nucleotides used to make RNA and DNA are full of energy. Most of their energy is stored in the two bonds between their two terminal phosphates. The nucleotides of DNA are called deoxyribonucleotides (abbreviated dATP, dTTP, dCTP and dGTP). The nucleotides of RNA are called ribonucleotides (abbreviated ATP, UTP, CTP, and GTP). While these nucleotides have different types of ribose sugar and nitrogenous bases, they all have the two terminal phosphate groups, and so they all have equivalent energies. All the energy in the nucleotide comes from the two terminal phosphates.

The energy used to make DNA and RNA polymers, through the formation of phosphodiester bonds, is supplied by the triphosphate nucleotides when they polymerize to form nucleic acids. It takes energy to form all chemical bonds, and the breaking of these chemical bonds will release energy. The triphosphate nucleotides are special because the bonds between the phosphates take more energy to form (~8 kilocalories/mole) as compared to the bonds between the phosphate and the ribose (~5 kilocalories/mole). Therefore these high energy bonds between the phosphates release more energy when they are broken.

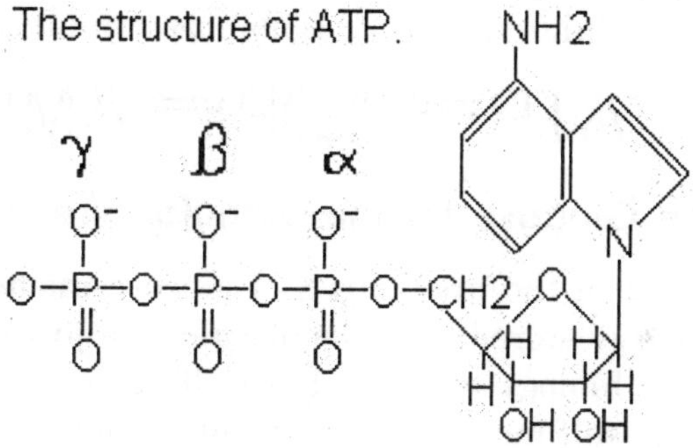

Figure : The structure of ATP

The phosphates in the structural formula of ATP have been assigned specific labels based on the IUPAC system. The phosphate attached to the #5 carbon of the ribose is labelled as the alpha (a) phosphate. The next two phosphates are labelled beta (ß) and gamma (g) from left to right. The bonds between the beta and gamma phosphates are represented by wavy lines which indicate the higher energy potential of these bonds.

Most biochemical reactions require energy input to proceed. These types of reactions are called endergonic reactions. An exergonic reaction, by contrast, gives off energy and proceeds spontaneously. Many chemical reactions of the cell use the energy from ATP which released when the beta or gamma phosphate bonds of ATP are broken. The energy released from the ATP supplies the energy necessary to form or break chemical bonds in biochemical reactions. The energy released by the conversion of ATP to ADP is used in the manufacture of ATP during the process of glycolysis.

When the phosphates are cleaved from ATP, two other forms of the nucleotide are made; ADP (adenosine diphosphate) and AMP (adenosine monophosphate). If you cover the terminal (gamma) phosphate in the structural formula of ATP with a picce of paper you will see what ADP looks like and if you cover the gamma and beta phosphates you will see what AMP looks like.

ATP is made inside cells by the transfer of a phosphate on to the b phosphate of ADP. Where does ADP come from? ADP is always inside cells as the by-product of the cleavage of a gamma phosphate from ATP. This is the ultimate use of recycling by the cell! Sometimes the ATP, ADP and AMP molecules are completely degraded because they wear out. Cells contain many biochemical pathways and can rebuild new AMP molecules from scratch when necessary. The AMP molecule can then be phosphorylated to form ADP and ATP. However, the cell prefers to conserve energy, whenever possible, by the dephosphorylation of ATP to ADP and AMP and the rephosphorylation of these back to ATP. The following reaction represents the phosphorylation and dephosphorylation reactions of ATP.

$$\text{ATP} \rightleftharpoons \text{ADP} \rightleftharpoons \text{AMP}$$

How Cells Use Glucose for the Manufacture of ATP

The first step of the breakdown of glucose inside the cell, to utilize the potential energy stored in this molecule, is a process called glycolysis. Any basic college biology text will have all of the steps of the process of glycolysis. In this self instructional I want to give you an overview of the process, in the hope that you will understand what is taking place and not just memorize the process.

Where does glycolysis take place?

Glycolysis takes place inside the cytoplasm of the cell. The cytoplasm is the aqueous based solution inside a cell that has a variety of molecules solubilized in it. The contents of the cytoplasm include proteins, carbohydrates, nucleic acids, salts and a whole host of other soluble molecules.

What are the parts of the cell that are not considered cytoplasm?

The organelles inside the cytoplasm are not part of the cytoplasm. Remember, most organelles are bound by lipid membranes and are not soluble in the cytoplasm. The contents of organelles are also isolated from the cytoplasm due to the hydrophobic nature of their membranes. This means that the substances solubilized inside organelles stay separate from the rest of the cytoplasm. The enzymes necessary for glycolysis are solubilized in the cytoplasm of the cell.

What does the process of glycolysis accomplish?

- Glycolysis uses the energy from two ATP to perform a series of enzymatic reactions resulting in the manufacture of four ATP.
- Glycolysis also generates two three-carbon compounds called pyruvate that still contain energy. The destination for pyruvate is the Citric Acid Cycle (also known as Krebs cycle) which occurs inside the matrix of the mitochondria.
- Glycolysis results in the addition of two hydrogens (electrons) onto the energy carrier nicotinamide adenine dinucleotide (NAD) to make NADH. The process of adding electrons to a substrate is called reduction. The destination for the NADH is the electron transport chain also found in the mitochondria.

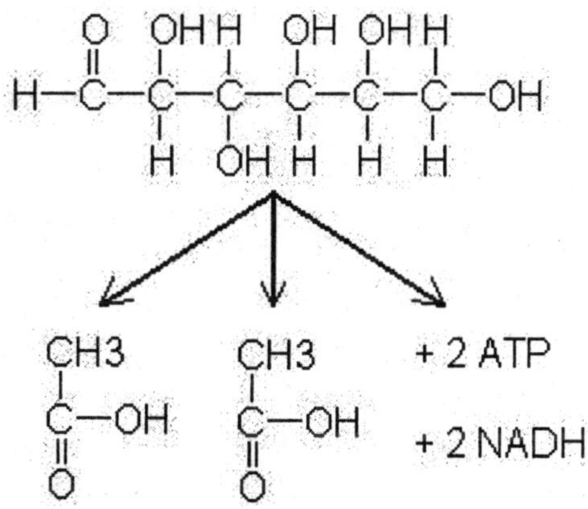

Figure : Splitting of glucose through glycolysis

Why are only two net ATP produced during glycolysis?

Although glycolysis allows for the formation of four ATP, two ATP are used in the process. This results in a net profit of two ATP from the process of Glycolysis.

Does glycolysis require molecular oxygen? No, glycolysis is an anaerobic process.

Enzymes and Enzymatic Reactions

To help understand the processes of glucose manufacture, anabolic reactions, and catabolic reactions, we need to understand how enzymes work.

Enzymes are a group of biological molecules, most of which are made of proteins. Enzymes are biological catalysts. A catalyst is added to a chemical reaction to speed the rate of the reaction. In other words, a catalyst makes the reaction occur faster. The catalyst is not changed in its form, but instead retains its original physical and chemical characteristics at the completion of the reaction. A catalyst does not participate in a reaction and can be reused again and again. Enzymes have all the qualities of a catalyst, but they are specifically used in the chemical reactions of living organisms (metabolism). Often people say; enzymes break this substance down or form this substance; however, to state the function of the enzyme correctly, one should say enzymes catalyse the formation or breakdown of a substance.

How do enzymes work? Enzymatic catalysts reduce the amount of energy necessary for endergonic reactions to proceed. Enzymes don't supply the energy, they reduce the amount needed. How? Well, there are several theories concerning this.

One way to speed reactions is to concentrate the materials that participate in them. Remember the generalized formula for chemical reactions?

Reactant A + Reactant B ———> Product C + Product D

Enzymes are molecules that have a 3-dimensional shape. Think of a muffin tin with all the depressions used to hold the individual muffins. If one hole contains reactant A and the adjacent hole contains reactant B, the two reactants are in close proximity to one another. This has the effect of concentrating the two reactants necessary for an anabolic reaction to occur and form products C and D.

Remember the cytoplasm of the cell or even the contents of an organelle may seem tiny, but it is a vast ocean of many components to a glucose or other small molecule.

Concentrating reactants is one way to speed reactions. Another way to reduce the energy and speed the rate of catabolic reactions might be to hold a molecule in place so that some of its bonds are under strain. Imagine you were trying to cut a thick piece of cloth with dull scissors and the going was very slow. If two people were to hold the material taut, or under strain, the fabric would cut more easily. Perhaps enzymes hold molecules in a similar fashion allowing bonds to be broken more easily.

If enzymes only speed the rates of reactions, does this mean that chemical reactions occur spontaneously anyway? Yes, all chemical reactions can occur spontaneously, but some reactions may take millennia to occur. Since organisms do not live on a geological time scale, the ability of enzymes to speed the rate of reactions, up to thousands of times faster, is necessary for life to exist.

The Structure of Enzymes

Enzymes are 3-dimensional molecules that have folded up into a specific shape (eg. the muffin tin). Imagine the depression in our muffin tin to be the site at which the chemical reaction takes place. This site is called the active site of the enzyme. The molecule that fits in the active site of the enzyme is called the substrate. Enzymes and substrates are very specific for each other. The active site of an enzyme is designed to fit perfectly with its specific substrate, like a 3-dimensional jigsaw puzzle.

Biochemical or Enzymatic Pathways

Sometimes a cell needs a product from a reactant that requires only a single chemical reaction. For example, reactant A needs to be changed to product B. This reaction would only require the enzyme that fits with its substrate, reactant A. However, this is not usually the case. Most products inside cells that need to be broken down or built up require a multi-step process. This process is called a biochemical pathway. It goes something like this:

- Enzyme A changes substrate A to product B.
- Product or substrate B now can be bound by enzyme B and be changed to product or substrate C
- Substrate C now fits with enzyme C which changes it to product D.
- If product D is what the cell needs, the pathway is finished. If not, the pathway can continue in the same way until the ultimate product is obtained.

Finally a quick lesson in naming enzymes. Generally to name an enzyme we take the substrate it breaks down or product that it manufactures and add an -ase on the end.

Examples: enzymes that degrade the following substances DNAse, ATPase or enzymes that build up the following substances: DNA polymerase and ATP synthetase..

Glucose Utilization

Both of the pyruvate (three-carbon molecules) are immediately converted to two-carbon molecules by removing a CO_2 (carbon dioxide) from each molecule. The remaining two-carbon compound is known as an acetyl group. The acetyl group will react quickly with another molecule because of its unpaired electrons. Therefore as the CO_2 is cleaved from pyruvate, the acetyl group is added simultaneously to a carrier molecule called coenzyme A (CoA).

Coenzyme A (CoA) is not an enzyme but a molecule similar in some respects to a nucleotide. This is a very important carrier molecule because it carries the acetyl group to many important biochemical pathways. The acetyl group and CoA molecule combine to form acetyl CoA. This molecule travels to the matrix of the mitochondria.

Once there, the CoA transfers the two-carbon acetyl group to an enzyme which catalyzes the addition of a four-carbon compound called oxaloacetic acid (oxaloacetate) to the acetyl. The result of this reaction is a six-carbon compound called citric acid (citrate). This is why this biochemical pathway is called the citric acid cycle.

The citric acid molecule has carbon, hydrogen, and oxygen atoms stripped from it in a step-wise fashion through this biochemical pathway. Each step requires its own specific enzyme. The end result of this pathway is the regeneration of oxaloacetic acid, and the conversion of NAD and FAD to NADH and FADH with the formation of carbon dioxide CO_2 as a by-product.

The major objective of the citric acid cycle is to get the hydrogens, along with their electrons, off the carbon compounds so they can be carried to the electron transport proteins that are embedded in the membrane of the mitochondrion. The waste or by-products of these reactions are the carbon and oxygen that are stripped off as each compound is rearranged by the removal of hydrogen. These carbons and oxygens combine to form CO_2 carbon dioxide gas which is given off as a by-product of these reactions. The CO_2 that you breathe out of your lungs is a waste product coming from the cellular metabolism of glucose!

Carbon dioxide (CO_2) enters our bloodstream, eventually winds up in our lungs, and is expelled upon exhalation, but where do those hydrogens with their electrons go? The hydrogens and electrons are picked up by electron carrier NAD (nicotinamide adenine dinucleotide) and electron carrier FAD (flavin adenine dinucleotide).

These electron carriers transfer the hydrogens and electrons to the electron carrying proteins, embedded in the mitochondrial membrane, which form the electron transport chain. The electron transport chain proteins are complexed with metal ions which allow them to carry extra electrons safely.

When NADH and FADH arrive at the electron transport chain, they transfer the electrons associated with the hydrogens to the electron carrying proteins at the top of the electron chain.

When electrons and hydrogen atoms are added to a molecule, this is called reduction. When electrons or hydrogens are removed from a molecule, this is known as oxidation. NAD is in an oxidized state while NADH is in a reduced state.

Now that the electrons have been transferred to the electron transport proteins, the hydrogen ion H+ (proton) is free to be pumped across the membrane into a space between the inner and outer mitochondrial membranes.

This proton pumping requires energy which is generated by the flow of electrons as they are passed down the electron transport protein chain. As electrons are transferred from one protein to the next, energy is released at each transfer, until the final electron acceptor, oxygen is reached. Aerobic organisms must take in oxygen because it serves as the final electron acceptor in the electron transport chain.

The energy, released from the passage of the electrons down the electron transport chain to oxygen, is used to pump the protons across the membrane. The pumping of protons requires a great deal of energy, because they have to be pumped against a concentration gradient.

Because protons are constantly being pumped across the membrane, they fill the space between the mitochondrial membranes. This is known as an area of high proton concentration. Putting more protons in this area would be like trying to force balloons into a box already filled to capacity with balloons.

Pushing protons against a concentration gradient of protons is referred to as going against, or up, a concentration gradient. In the case of the balloons, as soon as you open the box they would come out easily with no energy input necessary.

This is termed going with, or down, the concentration gradient. Molecules naturally flow from areas of high concentration to areas of low concentration with no energy input necessary.

Once you have captured all this energy by pushing protons up the proton concentration gradient into a crowded area, what happens? The protons can get out but they only have one passage to escape. This is a passage created by a protein enzyme called ATP synthetase. The ATP synthetase phosphorylates (adds phosphates to) ADP. The ATP synthetase gets the energy necessary for the formation of the high energy phosphate bond from the rush of protons through the channel it creates across the mitochondrial membrane.

To better understand how the flow of protons provides energy for the ATP synthetase enzyme imagine how a turbine engine works at Hoover Dam. The rushing water is caught and causes the huge turbine to turn which generates electricity. In a similar way the energy of protons cascading through the ATP synthetase is converted into the energy used to form the phosphate bond of ATP.

This process, called oxidative phosphorylation, makes most of the ATP manufactured by cells. While glycolysis gives us net profit of two ATP, the electron transport coupled with the citric acid cycle and oxidative phosphorylation give us a net profit of approximately 34 ATP from each glucose catabolized. Quite a big difference in yield!

What happens to the protons and electrons when they flow back through the membrane and down electron transport chain? Since electrons are energetic enough to cause problems inside the cell they must be inactivated. The final electron acceptor, oxygen, combines with the electrons and excess hydrogen ions to form water, a non-toxic by-product.

So the products of respiration are CO_2 (from the citric acid cycle) and H_2O (from electron transport) and lots of ATP. Without oxygen, aerobic cells die because the manufacture of ATP is halted at the cellular level. Since oxygen is absolutely necessary for the continual phosphorylation of ADP to ATP this process is called oxidative phosphorylation. Without oxygen, electron transport would stop, protons would not be pumped, and the ATP synthetase would not have a flow of protons to supply the energy necessary for the manufacture of ATP.

Do you know where all the molecules come from and their destination in the following Generalized Equation for Respiration?

$$C6H_{12}O_6 + O_2 \longrightarrow CO_2 + H_2O + \text{energy}$$

Now I would like to give you my overview of all of the processes of glucose metabolism in narrative form. I hope that you will understand the processes of glycolysis and respiration well enough to be able to describe this process in a similar way. In other words, I would like you to be able to think through and understand the concepts, not the minute details of these processes.

- Glucose is taken up as food by the body and carried to a cell. Glucose enters the cell by facilitated diffusion and is immediately broken down by the process of glycolysis in the cytoplasm of the cell.
- The process of glycolysis uses a biochemical pathway consisting of 10 enzyme mediated reactions which uses two ATP and manufactures four ATP (a net profit of two ATP). Glycolysis generates two NADH, which carry hydrogen and electrons to the electron transport chain in the mitochondrial membrane. Glycolysis also generates two three-carbon pyruvate molecules.
- The pyruvate molecule has a CO_2 removed and the remaining two-carbon acetyl group is then added onto the carrier molecule Coenzyme A to form acetyl CoA.
- Acetyl CoA is taken to the mitochondrial matrix to participate in the Citric acid cycle. The two-carbon acetyl group is then added via an enzyme mediated reaction to a four-carbon oxaloacetic acid molecule which results in the formation of the six-carbon molecule citric acid. Through a series of enzyme mediated reactions oxygens, carbons and hydrogens are systematically stripped off. The carbons

and oxygens combine in the appropriate proportions to form CO_2 and ultimately are expelled as waste products.

- The hydrogen atoms along with their electrons are picked up by the electron carriers NAD and FAD and taken to the electron transport proteins, which are located on the inner mitochondrial membrane.
- The electrons are handed off to the electron transport proteins where they flow from one protein to another releasing energy at each transfer until they combine with oxygen, the final electron acceptor in the electron transport chain.
- The energy released from the flow of electrons down the transport chain is used to pump protons across the inner mitochondrial membrane into the space between the inner and outer mitochondrial membranes, against a proton concentration gradient.
- The protons then flow back down the concentration gradient through an ATP synthetase enzyme channel, back into the mitochondrial matrix. The flow of protons through the ATP synthetase generates the energy necessary to phosphorylate ADP to ATP.
- Once back inside the matrix the protons combine with oxygen and electrons to form water a by-product of oxidative phosphorylation.

4

Enzymes Chemistry

Enzymes are proteins that catalyse (*i.e.*, increase or decrease the rates of) chemical reactions. In enzymatic reactions, the molecules at the beginning of the process are called substrates, and they are converted into different molecules, called the products. Almost all processes in a biological cell need enzymes to occur at significant rates. Since enzymes are selective for their substrates and speed up only a few reactions from among many possibilities, the set of enzymes made in a cell determines which metabolic pathways occur in that cell.

Like all catalysts, enzymes work by lowering the activation energy for a reaction, thus dramatically increasing the rate of the reaction. As a result, products are formed faster and reactions reach their equilibrium state more rapidly. Most enzyme reaction rates are millions of times faster than those of comparable un-catalysed reactions. As with all catalysts, enzymes are not consumed by the reactions they catalyse, nor do they alter the equilibrium of these reactions.

However, enzymes do differ from most other catalysts by being much more specific. Enzymes are known to catalyse about 4,000 biochemical reactions. A few RNA molecules called ribozymes also catalyse reactions, with an important example being some parts of the ribosome. Synthetic molecules called artificial enzymes also display enzyme-like catalysis.

Enzyme activity can be affected by other molecules. Inhibitors are molecules that decrease enzyme activity; activators are molecules that increase activity. Many drugs and poisons are enzyme inhibitors. Activity is also affected by temperature, chemical environment (*e.g.*, pH), and the concentration of substrate.

Some enzymes are used commercially, for example, in the synthesis of antibiotics. In addition, some household products use enzymes to speed up biochemical reactions (*e.g.*, enzymes in biological washing powders break down protein or fat stains on clothes; enzymes in meat tenderizers break down proteins, making the meat easier to chew).

Role of Enzymes in Biochemical Reactions

A living system controls its activity through enzymes. An enzyme is a protein molecule that is a biological catalyst with three characteristics. First, the basic function of an enzyme is to increase the rate of a reaction. Most cellular reactions occur about a million times faster than they would in the absence of an enzyme. Second, most enzymes act specifically with only one reactant (called a substrate) to produce products. The third and most remarkable characteristic is that enzymes are regulated from a state of low activity to high activity and vice versa. Gradually, you will appreciate that the individuality of a living cell is due in large part to the unique set of some 3,000 enzymes that it is genetically programmed to produce. If even one enzyme is missing or defective, the results can be disastrous.

Much of the information about enzymes has been made possible because they can be isolated from cells and made to work in a test tube environment. Extensive work has also been done with X-Ray diffraction techniques to elucidate the three-dimensional structure of some enzymes.

The ribbon and backbone form of carboxypeptidase is shown on the left. The substrate is shown in magenta.

Enzyme Parts List

The activity of an enzyme depends, at the minimum, on a specific protein chain. In many cases, the enzyme consists of the protein and a combination of one or more parts called cofactors. This enzyme complex is usually simply referred to simply as the enzyme.

Apoenzyme: The polypeptide or protein part of the enzyme is called the apoenzyme and may be inactive in its original synthesized structure. The inactive form of the apoenzyme is known as a proenzyme or zymogen. The proenzyme may contain several extra amino acids in the protein which are removed, and allows the final specific tertiary structure to be formed before it is activated as an apoenzyme.

Cofactors: A cofactor is a non-protein substance which may be organic, and called a coenzyme. The coenzyme is often derived from a vitamin with specific examples discussed later.

Another type of cofactor is an inorganic metal ion called a metal ion activator. The inorganic metal ions may be bonded through coordinate covalent bonds. The major reason for the nutritional requirement for minerals is to supply such metal ions as Zn^{+2}, Mg^{+2}, Mn^{+2}, Fe^{+2}, Cu^{+2}, K^{+1}, and Na^{+1} for use in enzymes as cofactors.

Final Enzyme: The type of association between the cofactor and the apoenzymes varies. In some cases, the bonds are rather loose and both come together only during the course of a reaction. In other cases, they are firmly bound together by covalent bonds. The activating role of a cofactor is to either: activate the protein by changing its geometric shape, or by actually participating in the overall reaction.

The overall enzyme contains a specific geometric shape called the active site where the reaction takes place. The molecule acted upon is called the substrate.

Enzyme Nomenclature and Classification

Enzymes are commonly named by adding a suffix "-ase" to the root name of the substrate molecule it is acting upon. For example, Lipase catalyzes the hydrolysis of a lipid triglyceride. Sucrase catalyzes the hydrolysis of sucrose into glucose and fructose.

A few enzymes discovered before this naming system was devised are known by common names. Examples are pepsin, trypsin, and chymotrypsin which catalyzes the hydrolysis of proteins.

The latest systematic nomenclature system known as the International Enzyme Commission (IEC) system is based upon the type of reaction catalysed. There are six broad groups of enzymes in this system as shown in table on the left.

For example, when using this system, "urease" becomes "urea amidohydrolase." Do not be overly concerned about enzyme names, but be able to recognize a substance as an enzyme by its "-ase" ending. Some types of reactions which are being catalysed will be self evident.

Structures and Mechanisms

Enzymes are generally globular proteins and range from just 62 amino acid residues in size, for the monomer of 4-oxalocrotonate tautomerase, to over 2,500 residues in the animal fatty acid synthase. A small number of RNA-based biological catalysts exist, with the most common being the ribosome; these are referred to as either RNA-enzymes or ribozymes. The activities of enzymes are determined by their three-dimensional structure. However, although structure does determine function, predicting a novel enzyme's activity just from its structure is a very difficult problem that has not yet been solved.

Most enzymes are much larger than the substrates they act on, and only a small portion of the enzyme (around 3–4 amino acids) is directly involved in catalysis. The region that contains these catalytic residues, binds the substrate, and then carries out the reaction is known as the active site. Enzymes can also contain sites that bind cofactors, which are needed for catalysis. Some enzymes also have binding sites for small molecules, which are often direct or indirect products or substrates of the reaction catalysed. This binding can serve to increase or decrease the enzyme's activity, providing a means for feedback regulation.

Like all proteins, enzymes are long, linear chains of amino acids that fold to produce a three-dimensional product. Each unique amino acid sequence produces a specific structure, which has unique properties. Individual protein chains may sometimes group together to form a protein complex. Most enzymes can be denatured—that is, unfolded and inactivated—by heating or chemical denaturants, which disrupt the three-dimensional structure of the protein. Depending on the enzyme, denaturation may be reversible or irreversible.

Structures of enzymes in complex with substrates or substrate analogs during a reaction may be obtained using Time resolved crystallography methods.

Specificity

Enzymes are usually very specific as to which reactions they catalyse and the substrates that are involved in these reactions. Complementary shape, charge and hydrophilic/hydrophobic characteristics of enzymes and substrates are responsible for this specificity. Enzymes can also show impressive levels of stereospecificity, regioselectivity and chemoselectivity.

Some of the enzymes showing the highest specificity and accuracy are involved in the copying and expression of the genome. These enzymes have "proof-reading" mechanisms. Here, an enzyme such as DNA polymerase catalyzes a reaction in a first step and then checks that the product is correct in a second step. This two-step process results in average error rates of less than 1 error in 100 million reactions in high-fidelity mammalian polymerases. Similar proofreading mechanisms are also found in RNA polymerase, aminoacyl tRNA synthetases and ribosomes.

Some enzymes that produce secondary metabolites are described as promiscuous, as they can act on a relatively broad range of different substrates. It has been suggested that this broad substrate specificity is important for the evolution of new biosynthetic pathways.

"Lock and Key" Model

Enzymes are very specific, and it was suggested by the Nobel laureate organic chemist Emil Fischer in 1894 that this was because both the enzyme and the substrate possess specific complementary geometric shapes that fit exactly into one another. This is often referred to as "the lock and key" model. However, while this model explains enzyme specificity, it fails to explain the stabilization of the transition state that enzymes achieve.

In 1958, Daniel Koshland suggested a modification to the lock and key model: since enzymes are rather flexible structures, the active site is continually reshaped by interactions with the substrate as the substrate interacts with the enzyme. As a result, the substrate does not simply bind to a rigid active site; the amino acid side chains which make up the active site are molded into the precise positions that enable the enzyme to perform its catalytic function. In some cases, such as glycosidases, the substrate molecule also changes shape slightly as it enters the active site. The active site continues to change until the substrate is completely bound, at which point the final shape and charge is determined. Induced fit may enhance the fidelity of molecular recognition in the presence of competition and noise via the conformational proofreading mechanism.

Mechanisms

Enzymes can act in several ways, all of which lower $\Delta G^{\ddagger}$:

- Lowering the activation energy by creating an environment in which the transition state is stabilized (e.g. straining the shape of a substrate—by binding the transition-state conformation of the substrate/product molecules, the enzyme distorts the bound substrate(s) into their transition state form, thereby reducing the amount of energy required to complete the transition).
- Lowering the energy of the transition state, but without distorting the substrate, by creating an environment with the opposite charge distribution to that of the transition state.
- Providing an alternative pathway. For example, temporarily reacting with the substrate to form an intermediate ES complex, which would be impossible in the absence of the enzyme.
- Reducing the reaction entropy change by bringing substrates together in the correct orientation to react. Considering $\Delta H^{\ddagger}$ alone overlooks this effect.
- Increases in temperatures speed up reactions. Thus, temperature increases help the enzyme function and develop the end product even faster. However, if heated too much, the enzyme's shape deteriorates and only when the temperature comes back to normal does the enzyme regain its shape. Some enzymes like thermolabile enzymes work best at low temperatures.

Interestingly, this entropic effect involves destabilization of the ground state, and its contribution to catalysis is relatively small.

Transition State Stabilization

The understanding of the origin of the reduction of $\Delta G^{\ddagger}$ requires one to find out how the enzymes can stabilize its transition state more than the transition state of the uncatalyzed reaction. Apparently, the most effective way for reaching large stabilization is the use of electrostatic effects, in particular, by having a relatively fixed polar environment that is oriented toward the charge distribution of the transition state. Such an environment does not exist in the uncatalyzed reaction in water.

Dynamics and Function

The internal dynamics of enzymes is linked to their mechanism of catalysis. Internal dynamics are the movement of parts of the enzyme's structure, such as individual amino acid residues, a group of amino acids, or even an entire protein domain. These movements occur at various time-scales ranging from femtoseconds to seconds. Networks of protein residues throughout an enzyme's structure can contribute to catalysis through dynamic motions.

Protein motions are vital to many enzymes, but whether small and fast vibrations, or larger and slower conformational movements are more important depends on the type of reaction involved. However, although these movements are important in binding and releasing substrates and products, it is not clear if protein movements help to accelerate the chemical steps in enzymatic reactions. These new insights also have implications in understanding allosteric effects and developing new drugs.

Allosteric Modulation

Allosteric sites are sites on the enzyme that bind to molecules in the cellular environment. The sites form weak, noncovalent bonds with these molecules, causing a change in the conformation of the enzyme. This change in conformation translates to the active site, which then affects the reaction rate of the enzyme. Allosteric interactions can both inhibit and activate enzymes and are a common way that enzymes are controlled in the body.

Cofactors and Coenzymes

Cofactors

Some enzymes do not need any additional components to show full activity. However, others require non-protein molecules called cofactors to be bound for activity. Cofactors can be either inorganic (*e.g.*, metal ions and iron-sulphur clusters) or organic compounds (e.g., flavin and heme). Organic cofactors can be either prosthetic groups, which are tightly bound to an enzyme, or coenzymes, which are released from the enzyme's active site during the reaction. Coenzymes include NADH, NADPH and adenosine triphosphate. These molecules transfer chemical groups between enzymes.

An example of an enzyme that contains a cofactor is carbonic anhydrase, and is shown in the ribbon diagram above with a zinc cofactor bound as part of its active site. These tightly bound molecules are usually found in the active site and are involved in catalysis. For example, flavin and heme cofactors are often involved in redox reactions.

Enzymes that require a cofactor but do not have one bound are called *apoenzymes* or *apoproteins*. An apoenzyme together with its cofactor(s) is called a *holoenzyme* (this is the active form). Most cofactors are not covalently attached to an enzyme, but are very tightly bound. However, organic prosthetic groups can be covalently bound (*e.g.*, thiamine pyrophosphate in the enzyme pyruvate dehydrogenase). The term "holoenzyme" can also be applied to enzymes that contain multiple protein subunits, such as the DNA polymerases; here the holoenzyme is the complete complex containing all the subunits needed for activity.

Coenzymes

Coenzymes are small organic molecules that transport chemical groups from one enzyme to another. Some of these chemicals such as riboflavin, thiamine and folic acid are vitamins (compounds which cannot be synthesized by the body and must be acquired from the diet). The chemical groups carried include the hydride ion (H^-) carried by NAD or $NADP^+$, the phosphate group carried by adenosine triphosphate, the acetyl group carried by coenzyme A, formyl, methenyl or methyl groups carried by folic acid and the methyl group carried by S-adenosylmethionine.

Since coenzymes are chemically changed as a consequence of enzyme action, it is useful to consider coenzymes to be a special class of substrates, or second substrates, which are common to many different enzymes. For example, about 700 enzymes are known to use the coenzyme NADH.

Coenzymes are usually continuously regenerated and their concentrations maintained at a steady level inside the cell: for example, NADPH is regenerated through the pentose phosphate pathway and *S*-adenosylmethionine by methionine adenosyltransferase. This continuous regeneration means that even small amounts of coenzymes are used very intensively. For example, the human body turns over its own weight in ATP each day.

Thermodynamics

As all catalysts, enzymes do not alter the position of the chemical equilibrium of the reaction. Usually, in the presence of an enzyme, the reaction runs in the same direction as it would without the enzyme, just more quickly. However, in the absence of the enzyme, other possible uncatalyzed, "spontaneous" reactions might lead to different products, because in those conditions this different product is formed faster.

Furthermore, enzymes can couple two or more reactions, so that a thermodynamically favourable reaction can be used to "drive" a thermodynamically unfavourable one. For example, the hydrolysis of ATP is often used to drive other chemical reactions.

Enzymes catalyse the forward and backward reactions equally. They do not alter the equilibrium itself, but only the speed at which it is reached. For example, carbonic anhydrase catalyzes its reaction in either direction depending on the concentration of its reactants.

$$CO_2 + H_2O \xrightarrow{\text{Carbonic anhydrase}} H_2CO_3$$

(in tissues; high CO_2 concentration)

$$H_2CO_3 \xrightarrow{\text{Carbonic anhydrase}} CO_2 + H_2O$$

(in lungs; low CO_2 concentration)

Nevertheless, if the equilibrium is greatly displaced in one direction, that is, in a very exergonic reaction, the reaction is *effectively* irreversible. Under these conditions the enzyme will, in fact, only catalyse the reaction in the thermodynamically allowed direction.

Kinetics

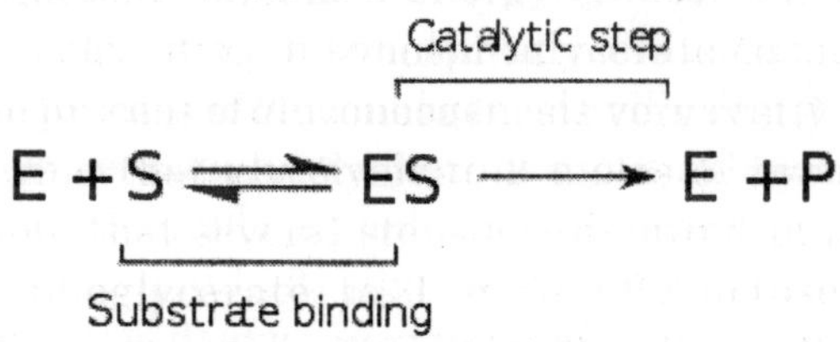

Mechanism for a single substrate enzyme catalysed reaction. The enzyme (E) binds a substrate (S) and produces a product (P).

Enzyme kinetics is the investigation of how enzymes bind substrates and turn them into products. The rate data used in kinetic analyses are obtained from enzyme assays.

In 1902 Victor Henri proposed a quantitative theory of enzyme kinetics, but his experimental data were not useful because the significance of the hydrogen ion concentration was not yet appreciated. After Peter Lauritz Sørensen had defined the logarithmic pH-scale and introduced the concept of buffering in 1909 the German chemist Leonor Michaelis and his Canadian postdoc Maud Leonora Menten repeated Henri's experiments and confirmed his equation which is referred to as Henri-Michaelis-Menten kinetics (sometimes also Michaelis-Menten kinetics). Their work was further developed by G. E. Briggs and J. B. S. Haldane, who derived kinetic equations that are still widely used today.

The major contribution of Henri was to think of enzyme reactions in two stages. In the first, the substrate binds reversibly to the enzyme, forming the enzyme-substrate complex. This is sometimes called the Michaelis complex. The enzyme then catalyzes the chemical step in the reaction and releases the product.

Enzymes can catalyse up to several million reactions per second. For example, the uncatalyzed decarboxylation of orotidine 5'-monophosphate has a half life of 78 million years. However, when the enzyme orotidine 5'-phosphate decarboxylase is added, the same process takes just 25 milliseconds. Enzyme rates depend on solution conditions and substrate concentration. Conditions that denature the protein abolish enzyme activity, such as high temperatures, extremes of pH or high salt concentrations, while raising substrate concentration tends to increase activity. To find the maximum speed of an enzymatic reaction, the substrate concentration is increased until a constant rate of product formation is seen.

This is shown in the saturation curve on the right. Saturation happens because, as substrate concentration increases, more and more of the free enzyme is converted into the substrate-bound ES form. At the maximum velocity (V_{max}) of the enzyme, all the enzyme active sites are bound to substrate, and the amount of ES complex is the same as the total amount of enzyme. However, V_{max} is only one kinetic constant of enzymes. The amount of substrate needed to achieve a given rate of reaction is also important. This is given by the Michaelis-Menten constant (K_m), which is the substrate concentration required for an enzyme to reach one-half its maximum velocity. Each enzyme has a characteristic K_m for a given substrate, and this can show how tight the binding of the substrate is to the enzyme. Another useful constant is k_{cat}, which is the number of substrate molecules handled by one active site per second.

The efficiency of an enzyme can be expressed in terms of k_{cat}/K_m. This is also called the specificity constant and incorporates the rate constants for all steps in the reaction. Because the specificity constant reflects both affinity and catalytic ability, it is useful for comparing different enzymes against each other, or the same enzyme with different substrates. The theoretical maximum for the specificity constant is called the diffusion limit and is about 10^8 to 10^9 ($M^{-1}\ s^{-1}$). At this point every collision of the enzyme with its substrate will result in catalysis, and the rate of product formation is not limited by the reaction rate but by the diffusion rate. Enzymes with this property are called *catalytically perfect* or *kinetically perfect*. Example of such enzymes are triose-phosphate isomerase, carbonic anhydrase, acetylcholinesterase, catalase, fumarase, β-lactamase, and superoxide dismutase.

Michaelis-Menten kinetics relies on the law of mass action, which is derived from the assumptions of free diffusion and thermodynamically driven random collision. However, many biochemical or cellular processes deviate significantly from these conditions, because of macromolecular crowding, phase-separation of the enzyme/substrate/product, or one or two-dimensional molecular movement. In these situations, a fractal Michaelis-Menten kinetics may be applied.

Some enzymes operate with kinetics which are faster than diffusion rates, which would seem to be impossible. Several mechanisms have been invoked to explain this phenomenon. Some proteins are believed to accelerate catalysis by drawing their substrate in and pre-orienting them by using dipolar electric fields.

Other models invoke a quantum-mechanical tunnelling explanation, whereby a proton or an electron can tunnel through activation barriers, although for proton tunnelling this model remains somewhat controversial. Quantum tunnelling for protons has been observed in tryptamine. This suggests that enzyme catalysis may be more accurately characterized as "through the barrier" rather than the traditional model, which requires substrates to go "over" a lowered energy barrier.

Inhibition

Enzyme reaction rates can be decreased by various types of enzyme inhibitors.

Competitive Inhibition

In competitive inhibition, the inhibitor and substrate compete for the enzyme (i.e., they can not bind at the same time). Often competitive inhibitors strongly resemble the real substrate of the enzyme. For example, methotrexate is a competitive inhibitor of the enzyme dihydrofolate reductase, which catalyzes the reduction of dihydrofolate to tetrahydrofolate.

Uncompetitive Inhibition

In uncompetitive inhibition the inhibitor can not bind to the free enzyme, but only to the ES-complex. The EIS-complex thus formed is enzymatically inactive. This type of inhibition is rare, but may occur in multimeric enzymes.

Non-competitive Inhibition

Non-competitive inhibitors can bind to the enzyme at the same time as the substrate, i.e. they *never* bind to the active site. Both the EI and EIS complexes are enzymatically inactive. Because the inhibitor can not be driven from the enzyme by higher substrate concentration (in contrast to competitive inhibition), the apparent V_{max} changes. But because the substrate can still bind to the enzyme, the K_m stays the same.

Mixed Inhibition

This type of inhibition resembles the non-competitive, except that the EIS-complex has residual enzymatic activity.

In many organisms inhibitors may act as part of a feedback mechanism. If an enzyme produces too much of one substance in the organism, that substance may act as an inhibitor for the enzyme at the beginning of the pathway that produces it, causing production of the substance to slow down or stop when there is sufficient amount. This is a form of negative feedback. Enzymes which are subject to this form of regulation are often multimeric and have allosteric binding sites for regulatory substances. Their substrate/velocity plots are not hyperbolar, but sigmoidal (S-shaped).

Irreversible inhibitors react with the enzyme and form a covalent adduct with the protein. The inactivation is irreversible. These compounds include eflornithine a drug used to treat the parasitic disease sleeping sickness. Penicillin and Aspirin also act in this manner. With these drugs, the compound is bound in the active site and the enzyme then converts the inhibitor into an activated form that reacts irreversibly with one or more amino acid residues.

Uses of Inhibitors

Since inhibitors modulate the function of enzymes they are often used as drugs. An common example of an inhibitor that is used as a drug is aspirin, which inhibits the COX-1 and COX-2 enzymes that produce the inflammation messenger prostaglandin, thus suppressing pain and inflammation. However, other enzyme inhibitors are poisons. For example, the poison cyanide is an irreversible enzyme inhibitor that combines with the copper and iron in the active site of the enzyme cytochrome c oxidase and blocks cellular respiration.

Biological Function

Enzymes serve a wide variety of functions inside living organisms. They are indispensable for signal transduction and cell regulation, often via kinases and phosphatases. They also generate movement, with myosin hydrolysing ATP to generate muscle contraction and also moving cargo around the cell as part of the cytoskeleton. Other ATPases in the cell membrane are ion pumps involved in active transport. Enzymes are also involved in more exotic functions, such as luciferase generating light in fireflies. Viruses can also contain enzymes for infecting cells, such as the HIV integrase and reverse transcriptase, or for viral release from cells, like the influenza virus neuraminidase.

An important function of enzymes is in the digestive systems of animals. Enzymes such as amylases and proteases break down large molecules (starch or proteins, respectively) into smaller ones, so they can be absorbed by the intestines. Starch molecules, for example, are too large to be absorbed from the intestine, but enzymes hydrolyse the starch chains into smaller molecules such as maltose and eventually glucose, which can then be absorbed.

Different enzymes digest different food substances. In ruminants which have herbivorous diets, microorganisms in the gut produce another enzyme, cellulase to break down the cellulose cell walls of plant fiber.

Several enzymes can work together in a specific order, creating metabolic pathways. In a metabolic pathway, one enzyme takes the product of another enzyme as a substrate. After the catalytic reaction, the product is then passed on to another enzyme. Sometimes more than one enzyme can catalyse the same reaction in parallel, this can allow more complex regulation: with for example a low constant activity being provided by one enzyme but an inducible high activity from a second enzyme.

Enzymes determine what steps occur in these pathways. Without enzymes, metabolism would neither progress through the same steps, nor be fast enough to serve the needs of the cell. Indeed, a metabolic pathway such as glycolysis could not exist independently of enzymes. Glucose, for example, can react directly with ATP to become phosphorylated at one or more of its carbons. In the absence of enzymes, this occurs so slowly as to be insignificant. However, if hexokinase is added, these slow reactions continue to take place except that phosphorylation at carbon 6 occurs so rapidly that if the mixture is tested a short time later, glucose-6-phosphate is found to be the only significant product. Consequently, the network of metabolic pathways within each cell depends on the set of functional enzymes that are present.

Control of activity

There are five main ways that enzyme activity is controlled in the cell.

1. Enzyme production (transcription and translation of enzyme genes) can be enhanced or diminished by a cell in response to changes in the cell's environment. This form of gene regulation is called enzyme induction and inhibition. For example, bacteria may become resistant to antibiotics such as penicillin because enzymes called beta-lactamases are induced that hydrolyse the crucial beta-lactam ring within the penicillin molecule. Another example are enzymes in the liver called cytochrome P450 oxidases, which are important in drug metabolism. Induction or inhibition of these enzymes can cause drug interactions.
2. Enzymes can be compartmentalized, with different metabolic pathways occurring in different cellular compartments. For example,

fatty acids are synthesized by one set of enzymes in the cytosol, endoplasmic reticulum and the Golgi apparatus and used by a different set of enzymes as a source of energy in the mitochondrion, through b-oxidation.

3. Enzymes can be regulated by inhibitors and activators. For example, the end product(s) of a metabolic pathway are often inhibitors for one of the first enzymes of the pathway (usually the first irreversible step, called *committed step*), thus regulating the amount of end product made by the pathways. Such a regulatory mechanism is called a negative feedback mechanism, because the amount of the end product produced is regulated by its own concentration. Negative feedback mechanism can effectively adjust the rate of synthesis of intermediate metabolites according to the demands of the cells. This helps allocate materials and energy economically, and prevents the manufacture of excess end products. The control of enzymatic action helps to maintain a stable internal environment in living organisms.
4. Enzymes can be regulated through post-translational modification. This can include phosphorylation, myristoylation and glycosylation. For example, in the response to insulin, the phosphorylation of multiple enzymes, including glycogen synthase, helps control the synthesis or degradation of glycogen and allows the cell to respond to changes in blood sugar. Another example of post-translational modification is the cleavage of the polypeptide chain. Chymotrypsin, a digestive protease, is produced in inactive form as chymotrypsinogen in the pancreas and transported in this form to the stomach where it is activated. This stops the enzyme from digesting the pancreas or other tissues before it enters the gut. This type of inactive precursor to an enzyme is known as a zymogen.
5. Some enzymes may become activated when localized to a different environment (e.g. from a reducing (cytoplasm) to an oxidizing (periplasm) environment, high pH to low pH etc.). For example, hemagglutinin in the influenza virus is activated by a conformational change caused by the acidic conditions, these occur when it is taken up inside its host cell and enters the lysosome.

Involvement in Disease

Since the tight control of enzyme activity is essential for homeostasis, any malfunction (mutation, overproduction, underproduction or deletion) of a single critical enzyme can lead to a genetic disease. The importance of enzymes is shown by the fact that a lethal illness can be caused by the malfunction of just one type of enzyme out of the thousands of types present in our bodies. One example is the most common type of phenylketonuria. A mutation of a single amino acid in the enzyme phenylalanine hydroxylase, which catalyzes the first step in the degradation of phenylalanine, results in build-up of phenylalanine and related products. This can lead to mental retardation if the disease is untreated.

Another example is when germline mutations in genes coding for DNA repair enzymes cause hereditary cancer syndromes such as xeroderma pigmentosum. Defects in these enzymes cause cancer since the body is less able to repair mutations in the genome. This causes a slow accumulation of mutations and results in the development of many types of cancer in the sufferer.

Naming Conventions

An enzyme's name is often derived from its substrate or the chemical reaction it catalyzes, with the word ending in-*ase*. Examples are lactase, alcohol dehydrogenase and DNA polymerase. This may result in different enzymes, called isozymes, with the same function having the same basic name. Isoenzymes have a different amino acid sequence and might be distinguished by their optimal pH, kinetic properties or immunologically. Furthermore, the normal physiological reaction an enzyme catalyzes may not be the same as under artificial conditions. This can result in the same enzyme being identified with two different names. *E.g.* Glucose isomerase, used industrially to convert glucose into the sweetener fructose, is a xylose isomerase *in vivo*.

The International Union of Biochemistry and Molecular Biology have developed a nomenclature for enzymes, the EC numbers; each enzyme is described by a sequence of four numbers preceded by "EC". The first number broadly classifies the enzyme based on its mechanism.

The top-level classification is:

- EC 1 *Oxidoreductases*: catalyse oxidation/reduction reactions
- EC 2 *Transferases*: transfer a functional group (*e.g.* a methyl or phosphate group)
- EC 3 *Hydrolases*: catalyse the hydrolysis of various bonds
- EC 4 *Lyases*: cleave various bonds by means other than hydrolysis and oxidation
- EC 5 *Isomerases*: catalyse isomerization changes within a single molecule
- EC 6 *Ligases*: join two molecules with covalent bonds.

The complete nomenclature can be browsed at http://www.chem.qmul.ac.uk/iubmb/enzyme/.

According to the naming conventions, enzymes are generally classified into six main family classes and many sub-family classes. Some web-servers, e.g., EzyPred and bioinformatics tools have been developed to predict which main family class and sub-family class an enzyme molecule belongs to according to its sequence information alone via the pseudo amino acid composition.

Industrial Applications

Enzymes are used in the chemical industry and other industrial applications when extremely specific catalysts are required. However, enzymes in general are limited in the number of reactions they have evolved to catalyse and also by their lack of stability in organic solvents and at high temperatures. Consequently, protein engineering is an active area of research and involves attempts to create new enzymes with novel properties, either through rational design or *in vitro* evolution. These efforts have begun to be successful, and a few enzymes have now been desiged "from scratch" to catalyse reactions that do not occur in nature.

Zinc Enzymes and Biological Catalysts

Zinc, also known as spelter, is a metallic chemical element; it has the symbol Zn and atomic number 30. It is the first element in group 12 of the periodic table. Zinc is, in some respects, chemically similar to magnesium, because its ion is of similar size and its only common oxidation state is +2. Zinc is the 24th most abundant element in the Earth's crust and has five stable isotopes. The most exploited zinc ore is sphalerite, a zinc sulfide. The largest exploitable deposits are found in Australia, Asia, and the United States. Zinc production includes froth flotation of the ore, roasting, and final extraction using electricity (electrowinning).

Brass, which is an alloy of copper and zinc, has been used since at least the 10th century BC. Impure zinc metal was not produced in large scale until the 13th century in India, while the metal was unknown to Europe until the end of the 16th century. Alchemists burned zinc in air to form what they called "philosopher's wool" or "white snow".

The element was probably named by the alchemist Paracelsus after the German word *Zinke.* German chemist Andreas Sigismund Marggraf is normally given credit for discovering pure metallic zinc in 1746. Work by Luigi Galvani and Alessandro Volta uncovered the electrochemical properties of zinc by 1800. Corrosion-resistant zinc plating of steel (hot-dip galvanizing) is the major application for zinc. Other applications are in batteries and alloys, such as brass. A variety of zinc compounds are commonly used, such as zinc carbonate and zinc gluconate (as dietary supplements), zinc chloride (in deodorants), zinc pyrithione (anti-dandruff shampoos), zinc sulfide (in luminescent paints), and zinc methyl or zinc diethyl in the organic laboratory.

Zinc is an essential mineral of "exceptional biologic and public health importance". Zinc deficiency affects about two billion people in the developing world and is associated with many diseases. In children it causes growth retardation, delayed sexual maturation, infection susceptibility, and diarrhea, contributing to the death of about 800,000 children worldwide per year. Enzymes with a zinc atom in the reactive centre are widespread in biochemistry, such as alcohol dehydrogenase in humans. Consumption of excess zinc can cause ataxia, lethargy and copper deficiency.

Biological Role

Zinc is an essential trace element, necessary for plants, animals, and microorganisms. Zinc is found in nearly 100 specific enzymes (other sources say 300), serves as structural ions in transcription factors and is stored and transferred in metallothioneins. It is "typically the second most abundant transition metal [sic] in organisms" after iron and it is the only metal which appears in all enzyme classes.

In proteins, Zn ions are often coordinated to the amino acid side chains of aspartic acid, glutamic acid, cysteine and histidine. The theoretical and computational description of this zinc binding in proteins (as well as that of other transition metals) is difficult. There are 2–4 grams of zinc distributed throughout the human body. Most zinc is in the brain, muscle, bones, kidney, and liver, with the highest concentrations in the prostate and parts of the eye. Semen is particularly rich in zinc, which is a key factor in prostate gland function and reproductive organ growth.

In humans, zinc plays "ubiquitous biological roles". It interacts with "a wide range of organic ligands", and has roles in the metabolism of RNA and DNA, signal transduction, and gene expression. It also regulates apoptosis. A 2006 study estimated that about 10% of human proteins (2800) potentially bind zinc, in addition to hundreds which transport and traffic zinc; a similar *in silico* study in the plant *Arabidopsis thaliana* found 2367 zinc-related proteins.

In the brain, zinc is stored in specific synaptic vesicles by glutamatergic neurons and can "modulate brain excitability". It plays a key role in synaptic plasticity and so in learning. However it has been called "the brain's dark horse" since it also can be a neurotoxin, suggesting zinc homeostasis plays a critical role in normal functioning of the brain and central nervous system.

Enzymes

Zinc is a good Lewis acid, making it a useful catalytic agent in hydroxylation and other enzymatic reactions. The metal also has a flexible coordination geometry, which allows proteins using it to rapidly shift conformations to perform biological reactions. Two examples of zinc-containing enzymes are carbonic anhydrase and carboxypeptidase, which are vital to the processes of carbon dioxide (CO_2) regulation and digestion of proteins, respectively.

In vertebrate blood, carbonic anhydrase converts CO_2 into bicarbonate and the same enzyme transforms the bicarbonate back into CO_2 for exhalation through the lungs. Without this enzyme, this conversion would occur about one million times slower at the normal blood pH of 7 or would require a pH of 10 or more.

The non-related β-carbonic anhydrase is required in plants for leaf formation, the synthesis of indole acetic acid (auxin) and anaerobic respiration (alcoholic fermentation).

Carboxypeptidase cleaves peptide linkages during digestion of proteins. A coordinate covalent bond is formed between the terminal peptide and a C=O group attached to zinc, which gives the carbon a positive charge. This helps to create a hydrophobic pocket on the enzyme near the zinc, which attracts the non-polar part of the protein being digested.

Other Proteins

Zinc serves a purely structural role in zinc fingers, twists and clusters. Zinc fingers form parts of some transcription factors, which are proteins that recognize DNA base sequences during the replication and transcription of DNA. Each of the nine or ten Zn^{2+} ions in a zinc finger helps maintain the finger's structure by coordinately binding to four amino acids in the transcription factor. The transcription factor wraps around the DNA helix and uses its fingers to accurately bind to the DNA sequence.

In blood plasma, zinc is bound to and transported by albumin (60%, low-affinity) and transferrin (10%). Since transferrin also transports iron, excessive iron reduces zinc absorption, and vice-versa. A similar reaction occurs with copper. The concentration of zinc in blood plasma stays relatively constant regardless of zinc intake. Cells in the salivary gland, prostate, immune system and intestine use zinc signalling as one way to communicate with other cells.

Zinc may be held in metallothionein reserves within microorganisms or in the intestines or liver of animals. Metallothionein in intestinal cells is capable of adjusting absorption of zinc by 15–40%. However, inadequate or excessive zinc intake can be harmful; excess zinc particularly impairs copper absorption because metallothionein absorbs both metals.

All Enzymes are Proteins

A molecule of protein that acts as a catalyst in biological system is called an enzyme. Enzymes control all the activities of the living system.

The three common characteristics of enzymes are as follows: First, the enzyme increases the rate of all biological reactions. Most of the cellular reactions that take place in presence of enzyme (catalyst) are faster (million times) than those reactions that would occur in the absence of an enzyme. Secondly, most of the enzymes act with only one specific reactant (i.e., substrate) and that leads to the formation of products. Third, enzymes regulate from a lower activity state to high activity state or, vice versa.

All Enzymes are Proteins: All enzymes are proteins that take part in chemical reactions as catalyst. They are globular proteins which may contain 62 to 2,500 amino acids. The three-dimensional structure of the enzyme determines activities of different enzymes. Almost all enzymes are of long and linear chains that are composed of many amino acids. These chains can fold and produce a product of three-dimensional structure.

Working of an Enzyme(Proteins)

An enzyme works according to the Lock and Key model. The active site of an enzyme and the substrate has the complementary shape for which they fit with each other to form complex, which is called an enzyme-substrate complex.

Now, the catalysis takes place and the substrate molecule gets change in shape, and then it gets broken down or can combine with some another molecule to form a new substance. The enzyme returns to its normal and is ready for another reaction.

Denaturation of Proteins (Enzymes)

It is known that all enzymes are protein. But on heating, or in the change of pH, or in the presence of chemical denaturants, these enzymes become unfolded and also inactivated. The process is called Denaturation of proteins. The process of Denaturation can be reversible or it can be irreversible. It depends of the nature of enzyme. Thus, the catalytic effect of any enzyme depends upon temperature and pH.

Metals in the Living System

Traces of metal is necessary for the human and animal system to function efficiently. Some of the important metals required are iron, copper, zinc, sodium and potassium.

- Iron metal is a constituent of haemoglobin in blood which carries oxygen to various parts of the body in humans and animals.
- Copper and zinc are constituents of enzymes which are the organic catalysts for various functions.
- Sodium and potassium are necessary for the electrical activities of the nerves and for the contraction of muscles.

- Presence of excessive amounts of metals are toxic and are harmful.
- Iron supplement is necessary for the formation of haemoglobin, whose difficiency causes anaemia. But excess of iron salts can cause a disease called siderosis.
- Excess of copper metal in the human body leads to a disease known as Wilsons disease. The symptoms of this disease are nervous disorders, tremors, difficulty in swallowing and still joints.

Enzymes are called biological catalysts which can enhance the rate of a chemical or biochemical reaction. It combines the reactants together and facilitates the reaction to take place fast and effectively. Enzymes are very important for our survival. Enzymes are involved in all biological processes and help the body to complete those processes in time.

Healthy living with our body happens with the help of many useful metabolic processes that are catalysed by various enzymes. Enzymes perform synthesizing, mixing, delivering or secreting of important ingredients that are used by the human body. They also eliminate several unwanted chemicals that are not required by the human body.

Importance of Enzymes in Health

Enzymes are protein molecules that are specific to a particular task. So they are able to bind to a molecule and catalyse a reaction. Most of these enzymes are synthesized from components of food which is not digested properly, water and other nutritional supplements, or from breakdown products of tissues. Enzymes take parts of food that enter in your stomach and change them by breaking down into vital basic nutrients that the body can absorb.

These basic nutrients are what the body utilizes to build, maintain and repair all its parts and their functions. Enzymes are essential to sustain life as it works together coordinating with vitamins and minerals and act as a catalyst to speed up chemical reaction. If plenty of live and active enzymes in the food are not present, the digestive system is not able to extract enough nutrients from it. Instead, the body depletes its own enzyme reserves to break down the food. Some of the food gets rotten in intestines and toxins are released into the blood stream. So without enough enzymes the body is left hungry, not nourished, without more energy and slowly deteriorating.

Sources of Enzymes

1. Without enzymes, ageing occurs more quickly, the body becomes prone to many diseases earlier in life, and also acquire physical impairments and mental retardation as our bodies degenerate.

2. Consuming raw fishes, meats, vegetables, fruits and specially fermented foods contains great amount of enzymes as cooked foods usually lack some important enzymes.
3. Fermented foods have more enzymes present in them called as vegetable enzyme.
4. When consuming boiled fish or roasted meat, try to add raw vegetables in diet. Raw vegetables contain rich vegetable enzymes which will help in digestion.

Introduction to Functions of Enzymes

Enzymes are "biological catalysts." A "Biological" substance means the substance in question is produced or is derived from some living organism. A "Catalyst" is a substance which increases the rate of a chemical reaction, but it neither gets changed nor destroyed by the chemical reaction that it accelerates.

Enzymes which are a subclass of catalysts are very specific in nature. Each enzyme can catalyse very selective chemical reactions and acts on very selective substances. An enzyme acts like a "key" which can "unlock" complex compounds. In order to function as a key an enzyme has a specific multi-dimensional shape that matches a specific section of the "substrate" referred as compound or substance which undergoes the change.

Classification of Enzymes

The enzymes are classified into the six groups according to the reactions which they catalyse

Oxidoreductases

These enzymes catalyse the oxidation and reduction processes that are always coupled in the cell.

Hydrolases

These enzymes are involved in the catalysis of chemical reactions in which the chemical bonds are broken off through the addition of a molecule of water.

Transferases

These enzymes are responsible for the catalysis of biochemical reactions during which the ions other than hydrogen are transferred between the reactants.

Lyases

These enzymes are responsible for catalysing reactions that result in the rearrangement of the molecular structures within the reactants.

Isomerases

Theses enzymes catalyse only isomerization reactions.

Ligases

These enzymes add molecular groups to a substrate and hence catalyse the formation of a chemical bond between two unrelated molecules.

Some Specific Functions of Enzymes

Digestive enzymes: During the process of digestion, complex substances made up of protein, fats and carbohydrates are degraded down into basic nutrient materials that can be used by the human body. The proteolytic, lipolytic, and amylolytic enzymes work in tandem facilitating the break down of food into its basic constituent for easy assimilation by the cells.

Restriction enzymes: They are bacterial enzymes which cleave DNA at very specific locations.

Cardiac enzymes: These enzymes are capable of speeding up certain biochemical processes in the cardiac muscle. Abnormal levels of these enzymes result into heart attack.

Metabolic enzymes: These are protein-based substances that promote metabolic changes in the cells.

Effect of Catalyst

Everyday experience tells us that chemical reactions take place in widely varying rates. Infact rates of various reactions vary from very slow to very fast. Some reactions are so rapid that they occur as soon as the reactants are mixed. For example, the reactions involving ionic species (known as ionic solutions) are very fast.

Common examples of such type are:

i) Acid-Base Neutralisation

$$H^+_{(aq)} + Cl^-_{(aq)} + Na^+_{(aq)} + OH^-_{(aq)} \longrightarrow H_2O(l) + Na^+ Cl^-_{(aq)}$$

ii) Precipitation from aqueous solution of NaCl and $AgNO_3$

$$Ag^+_{(aq)} + NO^-_{3(aq)} + Na^+_{(aq)} + Cl^-_{(aq)} \longrightarrow AgCl_{(S)} + Na^+_{(aq)} + NO_3^-{}_{(aq)}$$

iii) Ionic Reactions.

These reactions occur in about 10^{-14} to 10^{-16} seconds and are very fast. These reactions involve only the ions and no bonds are broken. Therefore their rates are very fast.

On the other hand there are certain reactions which occur at an extremely slow speed.

Examples of very slow reactions are:

- Rusting of iron
- Conversion of monoclinic sulphur to rhombic sulphur etc.

In between very slow and very fast reactions, we have reactions which proceed at a moderate speed which can easily be measured.

Examples of moderate speed reactions are:

i) The reaction between nitrogen peroxide and carbon monoxide

$$NO_{2(g)} + CO_{(g)} \longrightarrow CO_{2(g)} + NO_{(g)}$$

ii) Decomposition of hydrogen peroxide

$$2H_2O_{2(l)} \longrightarrow 2H_2O_{(l)} + O_{2(g)}$$

iii) Decomposition of dinitrogen pentoxide.

$$2N_2O_{5(g)} \longrightarrow 2N_2O_{4(g)} + O_{2(g)}$$

Rate of a Reaction

The rate of a reaction is broadly defined as the speed with which the reactants are converted into products. This can be determined both quantitatively and qualitatively. Qualitatively, (concerned with or depending on constitiuents) an idea about the rate of a reaction can be obtained by observing either the speed of disappearance of the reactants or the speed of appearance of the products. For example when a piece of magnesium is put into a beaker containing dilute hydrochloric acid, hydrogen gas is rapidly evolved and magnesium disappears at once. On the other hand, when a piece of iron is placed in the same acid, hydrogen gas evolves at a slower rate and iron also disappears very slowly. This means that the reaction of magnesium and hydrochloric acid is faster than reaction of iron and hydrochloric acid.

$$Mg_{(g)} + 2HCl_{(aq)} \longrightarrow MgCl_{2(aq)} + H_{2(g)} \text{(fast)}$$

$$Fe_{(g)} + 2HCl_{(aq)} \longrightarrow FeCl_{2(aq)} + H_{2(g)} \text{(slow)}$$

Units of Rate of Reaction

Units of concentration (in aqueous solutions) is moles per litre

Unit of time is seconds, minutes etc.

$$\therefore \text{Unit of rate of reaction} = \frac{\text{moles/litre}}{\text{second}} = \text{mol } L^{-1}s^{-1}$$

For e.g., Consider a hypothetical reaction

$$A \longrightarrow B$$

The rate of a reaction may be expressed in either of the following two ways:

1) The rate of disappearance or decrease in concentration of A (reactants)

$$\text{Rate of reaction} = \frac{\text{Decrease in concentration of A}}{\text{Time taken}}$$

2) The rate of appearance or increase in concentration of B (products)

$$\text{Rate of reaction} = \frac{\text{Increase in concentration of B}}{\text{Time taken}}$$

Equation A B shows that one mole of A produces one mole of B and thus the rate at which the concentration of A decreases will be the same as the rate at which the concentration of B increases.

The decrease or increase in the concentration of the reactants or products may also be expressed in terms of change in their concentration during time interval Dt as

$$\text{Rate of a reaction} = -\frac{\Delta[A]}{\Delta t}$$

or

$$\text{Rate of a reaction} = (+)\frac{\Delta[B]}{\Delta t}$$

Where D[A] gives the decrease in concentration of A and D[B] represents the increase in concentration of B.

The square brackets around the substances are used to express the concentration (in terms of mol/litre). It may be noted that in the case of concentration of reactants minus sign is used.

This implies that the concentration of the reactants is decreasing with time.

For e.g., consider the reaction:

$$NO + O_3 \longrightarrow NO_2 + O_2$$

The rate of the reaction can be expressed either by decrease of concentration of any of the reactants (NO or O_3) or by increase in concentration of any of the products (NO_2 or O_2). Thus,

$$\text{Rate of a reaction} = -\frac{\Delta [NO]}{\Delta t} = -\frac{\Delta [O_3]}{\Delta t}$$

or

$$\text{Rate of a reaction} = \frac{\Delta [NO_2]}{\Delta t} = \frac{\Delta [O_2]}{\Delta t}$$

However for the reaction

$$H_{2(g)} + I_{2(g)} \Leftrightarrow 2HI_{(g)}$$

We observe from the stoichiometry of the reaction, that one mole of H_2 reacts with one mole of I_2 to form two moles by hydrogen iodide. This means that the rate of disappearance of H_2 and I_2 is the same but the rate of appearance of HI must be twice the rate of disappearance of H_2 and I_2.

Thus we may write,

$$-2\frac{\Delta[H_2]}{\Delta t} = -2\frac{\Delta[I_2]}{\Delta t} = \frac{\Delta[HI]}{\Delta t}$$

But for having unique value of the reaction rate (independent of the concentration term chosen), we divide the rate of the reaction, defined with respect to any of the reactants or products by the number of moles involved in the reaction. Thus, for the above reaction, rate may be expressed as (by dividing by 2).

$$\text{Rate of reaction} = -\frac{\Delta[H_2]}{\Delta t}$$

$$= -\frac{\Delta[I_2]}{\Delta t}$$

$$= \frac{1}{2}\frac{[HI]}{\Delta t}$$

Consider another example,

When acidified hydrogen peroxide (H_2O_2) is added to a solution of potassium iodide (KI), iodine is liberated.

$$H_2O_{2(aq)} + 2KI_{(aq)} + H_2SO_{4(aq)} \rightarrow 2H_2O(l) + I_{2(aq)} + K_2SO_{4(aq)}$$

Concentration of iodine is zero initially. With the passage of time, the concentration of iodine increases and the reaction solution becomes brownish.

Concentration of iodine can be measured at different intervals of time by titrating against sodium thiosulphate solution.

If the concentration of iodine rises from 0 to 10^{-5} mol L^{-1} in seconds,

$$\text{Re action rate} = \frac{\Delta[I_2]}{\Delta t} = \frac{10^{-5}\,\text{mol}\,L^{-1}}{10\text{s}}$$

$$= 10^{-6}\ \text{mol}\,L^{-1}s^{-1}$$

[Here symbol D represents a change and I_2 represents the molar concentration of iodine].

This change in concentration of the product (iodine) takes place in ten seconds. Thus the rate can be called the average reaction rate.

The change in concentration over a shorter time interval will give a more accurate estimate of reaction at any moment.

Rate of reaction depends on the concentration of the reactants. With lapse of time, concentration of reactant/(s) decreases and therefore, rate of reaction also decreases.

In order to determine the rate of a particular instant, one should make the time interval almost zero.

$$\text{i.e.,}\ \frac{\Delta[I_2]}{\Delta t} = \frac{d[I_2]}{dt}$$

$$\Delta \rightarrow 0$$

$d[I_2]$ denotes a very small change in concentration of I_2, and dt denotes very small change in time.

$$\text{Thus rate of reaction} = \frac{d[I_2]}{dt}$$

This rate is instantaneous rate.

The concentration of H_2O_2 decreases with time and thus rate in terms of H_2O_2 and KI is expressed as

$$\text{Rate of reaction} = \frac{-\Delta[H_2O_2]}{\Delta t} = -\frac{1}{2}\frac{\Delta[KI]}{\Delta t}$$

The negative sign indicates that the concentration of H_2O_2 and KI decreases with time.

In this above reaction, rate of consumption of KI is twice the rate of

Thus to make the two rates equal we divide by two and write

$$-\frac{1}{2}\frac{\Delta[KI]}{\Delta t}$$

Consider another example,

Cisplatin, a compound with a molecular formula of Pt $(NH_3)_2$ Cl_2 is used in chemotherapy of cancer. It reacts with water and releases chloride ion bound to central platinum metal.

The reaction is represented as:

$$Pt(NH_3)_2Cl_2 + H_2O \longrightarrow Pt(NH_3)_2Cl^+ + Cl^-$$

The rate at which this reaction can occur is found by measuring the amounts of Pt $(NH_3)_2$ Cl_2 consumed or the amount of Cl^- released per unit volume of the reaction mixture.

The amount of cisplatin per unit volume at different intervals of time is obtained and provided in the given table.

Time/minutes	*Concentration of Cisplatin/mol L- 1*
0	0.0100
20	0.00970
40	0.00940
80	0.00887

$$= \frac{[Pt(NH_3)_2Cl_2] \text{ at } t_2 - [Pt(NH_3)_2Cl_2] \text{ at } t_1}{t_2 - t_1}$$

If we consider concentrations at $t_1 = 0$ min and $t_2 = 40$ min, then the rate of change of concentration of Cisplatin is calculated as follows:

Rate of change of concentration of Cisplatin

$$= \frac{(0.0094 - 0.0100)\text{ mol L}^{-1}}{(40 - 0)\text{ min}}$$

$$= \frac{-0.0006\text{ mol L}^{-1}}{40\text{ min}}$$

$$= -1.5 \times 10^{-5}\text{ mol L}^{-1}\text{ min}^{-1}$$

The reaction rate is defined as the rate of decrease of concentration of a reactant, or the rate of increase of concentration of a product and is always given as a positive number.

When rate is expressed in terms of $\frac{\Delta[\text{reactant}]}{\Delta t}$ which is a negative quantity, a minus sign is always used to make the rate positive.

Therefore the average rate of reaction between $t_1 = 0$ minutes and $t_2 = 40$ minutes will be

$$-\left[-1.5 \times 10^{-5} \text{ mol L}^{-1} \text{ min}^{-1}\right] = 1.5 \times 10^{-5} \text{ mol L}^{-1} \text{ min}^{-1}$$

Factors Affecting the Rate of a Reaction

The rate of reaction depends on the following factors:

- Concentration of the reactants
- Surface area of the reactants
- Temperature
- Catalyst
- Light.

Effect of Concentration

The rate of a reaction generally increases with increase in concentration of the reactants. For a reaction to take place between two molecules, they must collide or come into contact. The probability of such collisions increase with the increase in the concentration of the reacting molecules. Different reactants can affect the rate of a particular reaction to different extent. For example, consider the reaction

$$2NO + O_2 \longrightarrow 2NO_2$$

The rate of this reaction doubles when concentration of oxygen is doubled. On the other hand, when the concentration of nitric oxide is doubled, the rate of reaction increases four times. In systems where the reactants and the products are in different states, the area of contact between the reacting substances influences the rate considerably. It is found that magnesium powder reacts much more rapidly than magnesium ribbon with dilute H_2SO_4.

Activity to Study the Effect of Concentration

Requirements: 6g of granular zinc

10 mL of 1 M HCl, 10 mL of 2 M HCl, A graduated syringe and a boiling tube.

The syringe is attached to the boiling tube and the reaction is carried out with 3g of zinc and 5 mL of 1 M HCl and next with 3g of zinc and 5 mL of 2 M HCl.

The volume of the gas at time intervals of 20s, 40s and 60s is recorded and the results are plotted.

Effect of Surface Area of the Reactants

The rate of the reaction increases with increase in surface area of the reactants.

It is a common observation that small pieces of wood burn more rapidly than large logs of wood.

As mentioned earlier, magnesium powder reacts with dilute sulphuric acid more rapidly than magnesium ribbons. Thus smaller the size of reacting particles, the greater is the total surface area exposed for reaction and consequently greater is the rate of the reaction.

Effect of Temperature

Generally, the rates of chemical reactions increase with increase in temperature. It is a common observation that perishable foods like milk, cooked vegetables decay more quickly in summers than in winter.

Consider the chemical reaction:

$$CaCO_{3(s)} + 2HCl_{(aq)} \Leftrightarrow CaCl_{2(aq)} + CO_{2(g)} + H_2O_{(l)}$$

At 273 K, the reaction is extremely slow and the amount of CO_2 collected is very small. When the temperature is raised to 323 K (50° C) the evolution of CO_2 is more rapid.

The reaction between zinc and hydrochloric acid can be performed at two different temperatures: 293 K and 308 K.

It will be found that the volume of hydrogen formed during the same time interval will be more at 308K then at 298 K.

Effect of Catalyst

Catalysts are substances, which alter the rate of chemical reactions without undergoing any overall chemical change themselves.

In the manufacture of ammonia, iron is used as a catalyst and it enhances the rate of this reaction.

$$N_{2(g)} + 3H_{2(g)} \xrightarrow{\text{Fe(catalyst)}} 2NH_{3(g)}$$

In the reaction $2KClO_3 \rightarrow 2KCl + 3O_2$, manganese dioxide act as a catalyst.

Sometimes catalysts are used to retard or slow down the rates of reaction. For example., glycerol is used to slow down the decomposition of hydrogen peroxide. Such catalysts are known as negative catalysts.

Many industrial processes like the manufacture of polythene and polystyrene require catalysts.

In the living bodies, a large number of complex chemical reactions occur which are catalysed by complex organic molecules called enzymes. For example., enzyme amylase present in saliva catalyses the decomposition of starch into maltose.

Effect of Light

There are many reactions, which are influenced by light. For example., photosynthesis and photography. Such reactions are known as photochemical reactions.

Other examples are:

$$H_2 + Cl_2 \xrightarrow{\text{sunlight}} 2HCl$$

$$2AgCl \xrightarrow{\text{light}} 2Ag + Cl_2$$

$$SO_2 + Cl_2 \xrightarrow{\text{sunlight}} \underset{\text{sulphuryl chloride}}{SO_2Cl_2}$$

$$CO + Cl_2 \xrightarrow{\text{sunlight}} \underset{\text{phosgene}}{COCl_2}$$

Light supplies the necessary energy for the reaction to take place. Photographic films which are normally coated with silver bromide (and a very small amount of silver iodide) undergo chemical reaction when exposed to sunlight. This basic reaction is used in photography.

Enzyme Catalysis

Enzymes are macromolecules, usually proteins, produced in living systems, which act as catalysts in physiological reactions.

The striking characteristics of enzymes are their catalytic power and specificity. Enzymes have immense catalytic power; they accelerate reactions by factors of at least a million. Most reactions in living systems do not occur at perceptible rates in the absence of enzymes.

A simple reaction like hydration of CO_2 is catalysed by the enzyme carbonic anhydrase. The transfer of CO_2 from tissues into the blood and then to the alveolar air would be very slow in the absence of this enzyme. The enzyme can hydrate 10^5 molecules of CO_2 per second, which is 10^7 times faster than the unanalyzed one.

This example illustrates the fact that enzymes have immense catalytic power; enzymes increase rates of reaction by 10^6 to 10^{20} times. The catalytic power of proteins comes from their capacity to bind substrates (reactant) molecules in precise orientations and to stabilize transition states in the making and breaking of bonds.

Enzymes are highly specific. An enzyme usually catalyses a single chemical reaction or a set of closely related reactions. For e.g., urease catalyses the hydrolysis of urea only.

$$NH_2CONH_2 + H_2O \xrightarrow{\text{Urease}} 2NH_3 + 2CO_2$$

Other examples of enzyme catalysed reactions are as follows:

$$\underset{\text{Starch}}{2(C_6H_{10}O_5)x} + xH_2O \xrightarrow{\text{Diastase}} \underset{\text{Maltose}}{x\,C_{12}H_{22}O_{11}}$$

$$C_{12}H_{22}O_{11} + H_2O \xrightarrow{\text{Maltase}} \underset{\text{Glucose}}{2C_6H_{12}O_6}$$

$$\underset{\text{Glucose}}{C_6H_{12}O_6} \xrightarrow{\text{Zymase}} 2C_2H_5OH + 2CO_2$$

The remarkable specificity of enzymes comes from the ability of the enzymes to bring the substrates into a favourable orientation in the enzyme-substrate (ES) complexes such that the substrates are bound to specific regions of the enzymes. Each enzyme has a specific active site. The active site of an enzyme binds the substrate and contains the residues that directly participate in the making and breaking of bonds. Some key features of the active site are:

1) The active site takes up a relatively small part of the total volume of an enzyme.
2) The active site is a three dimensional entity formed by groups that come from different parts of the linear amino acid sequence.
3) Enzymes bind the substrate by multiple weak interactions. These weak interactions are electrostatic bonds, hydrogen bonds, van der Waals' forces and hydrophobic interactions. van der Waals' forces become significant in binding only when many substrate species can come simultaneously close to many enzymes species. Hence, the enzyme and substrate should have complementary shapes. The directional nature of hydrogen bonds between substrate and enzyme brings about a high degree of specificity.

4) Active sites are clefts and crevices. Substrate molecules bind to the clefts or crevices.

5) The specificity of binding of a substrate depends on the precise defined arrangement of atoms in the active site. To fit into the active site the substrate must have a matching shape, just like how a key fits into a lock.

This model was proposed in the late 19th century. Since then, modern X-ray crystallographic and spectroscopic methods have revealed that in many cases, enzymes are markedly modified by the binding of substrate. The active sites of these enzymes have shapes that are complementary to that of the substrate only after the substrate is bound. Such enzyme-substrate interaction are described by induced fit model.

In enzyme catalysed reaction, the rate of reaction varies with the concentration of the substrate in the manner.

At a given concentration of enzyme, the rate of reaction is almost linearly proportional to [S] when [S] is low. At high [S], r is nearly independent of [S]. To account for such behaviour, Leonor Michaelis and Maud Menten proposed the following steps in enzyme catalysed reactions.

$$E + S \underset{k_2}{\overset{k_1}{\rightleftharpoons}} ES \xrightarrow[(2)]{k_3} E + P$$
$$(1)$$

The formation of ES complex is fast and reversible, while the formation of the product (step 2) is the slow, rate determining step. The rate of enzyme catalysed reaction changes from first order to zero order as the concentration of the substrate is increased.

Enzymes are complex biochemical catalysts, speeding up a particular reaction to produce an ordered, stable reaction system in which the products of any reaction are made when they are needed. A specific enzyme controls each reaction in a series of metabolic reactions. Enzymes also control cell metabolism by regulating how and when reactions occur.

They are made up of globular proteins that have complex tertiary or quaternary structure. Enzyme shape is maintained by hydrogen bonds and ionic forces and their function can be affected by changes in temperature and pH.

Enzyme Catalysed Reaction has some Important Properties:

- Each enzyme is specific and catalyzes only one reaction at a time
- Enzymes combine with their substrates to form temporary enzyme-substrate complex.

- Enzymes are not altered or used up in the reactions they catalyse, so can be used again and again.
- Enzyme catalysed reactions are sensitive to temperature and pH.
- Enzyme catalysed reactions can be slowed down or stopped by inhibitors.
- Enzymes lower the activation energy of a reaction thus making the reaction to occur very rapidly with large turnover numbers.

Enzymes are Classified According to the Type of Reactions they Catalyse:

- Oxidoreductase: these catalyse oxidation and reduction reactions.
- Transferases: these catalyse the transfer of a chemical group from one compound to another, such as transfer of an amino group from one amino acid to another organic acid.
- Hydrolases: these catalyse hydrolysis reactions. Most digestive enzymes are hydrolases
- Lyases: these catalyse the breakdown of molecules by reactions that do not involve hydrolysis.
- Isomerases: these catalyse the transformation of one isomer into another.
- Ligases: these catalyse the formation of bonds between compounds often using the free energy made available from ATP hydrolysis. Example- DNA ligases.

Activation Energy is Lowered in Enzyme Catalysed Reactions

For any chemical reaction to take place, old bonds must be broken before new ones can form. The energy needed to break these bonds is called as activation energy. Many reactants need a high amount of energy to reach a state where they can take part in a reaction. In the presence of enzymes the activation energy is greatly lowered and this allows a reaction to take place at low temperature compared to a reaction without involvement of enzymes.

For example: When carbon dioxide dissolves in water, a small proportion of the CO_2 combines with H_2O to form carbonic acid. In the presence of enzyme carbonic anhydrase, it is speeded up around 10^7 times! One molecule of carbonic anhydrase can convert 600000 molecules of CO_2 every second.

5

Plant Molecular Physiology

Plant Cells

The Plant Kingdom

Since the time of Aristotle (384–322 B.C.E.), biologists have sought to classify organisms. At first the purpose was ease of identification ("artificial" classification schemes). Carolus Linnaeus (1707–1778), arguably the greatest of the pre-modern Naturalists, sought to classify plants and other organisms according to affinity groups that reflected the mind of the Creator. Later, after Darwin, the goal of classification was to show evolutionary relationships ("natural" classification schemes).

For the past 150 years, biologists have emphasized natural systems of classification and have attempted to define morphological criteria that reveal evolutionary relationships.

We now know that morphology, the form and structure of organisms, is the end product of the actions of genes. Virtually all of the information needed to form a complete organism is encoded in its DNA sequences, both nuclear and cytoplasmic (mitochondria and chloroplasts). DNA sequence analysis has thus provided evolutionary biologists with a powerful new tool for arriving at a truly natural classification system.

On the basis of phylogenetic analyses of highly conserved DNA sequences, living organisms have been divided into three major domains: Bacteria, Archaea, and Eucarya.

1. The common ancestor of all the organisms first gave rise to the Bacteria and the common ancestor of the Archaea and the Eucarya.
2. The Archaea branch off from the Eucarya lineage.
3. The Eucarya common ancestor acquires mitochondrial endosymbiont.

4. A heterogeneous group of eukaryotes called protists branch off the lineage leading to plants, fungi, and animals.
5. The common ancestor of fungi and animals form a branch, followed by a divergence into the fungal and animal lineages.
6. The common ancestor of plants, green algae, red algae, and glaucophytes acquires chloroplast endosymbiont (a cyanobacterium).
7. The three lineages of glaucophytes, red algae, and green algae diverge.
8. Various lineages of protists acquire chloroplasts via green or red algal endosymbionts.
9. The earliest branch of green algae diverge.
10. The later branch of green algae diverge.
11. The remaining lineage leads to plants.

The Eucarya include the eukaryotes, organisms whose cells contain a true nucleus. The Bacteria, or eubacteria, which include the cyanobacteria, lack a true nucleus and are therefore prokaryotic.

The Archaea, or archaebacteria, are also prokaryotic, but they differ from the Bacteria:

Besides their morphological and biochemical differences, they are often adapted to extreme environments, such as sulphur hot springs or saline ponds. Phylogenetic studies have indicated that the Archaea and Eucarya split after the Bacteria separated from the common ancestor. Thus Archaea and Eucarya represent sister groups. This closer relation between Archaea and Eucarya is reflected in their similar promoter structures and RNA polymerases, the presence of histones, and many other characteristics.

Fungi were formerly classified as algal-like plants that had lost their chloroplasts. However, fungi and animals branched off from the Eucarya lineage *before* the appearance of choloroplasts. They are thus more closely related to animals than to plants. Fungi are heterotrophic; that is, they depend on other organisms for their food, and they satisfy their nutritional needs by absorbing inorganic ions and organic molecules from the external environment. Most fungal species are filamentous and possess cell walls made of chitin, the same substance that is found in insect exoskeletons.

Bryophytes are small (rarely more than 4 cm in height), very simple land plants, and the least abundant in terms of number of species and overall population. Bryophytes do not appear to be in the direct line of evolution leading to the vascular plants; rather, they seem to constitute a separate minor branch.

Bryophytes include mosses, liverworts, and hornworts. These small plants have life cycles that depend on water during the sexual phase. Water facilitates fertilization, the fusion of gametes to produce a diploid zygote, a feature also seen in the algal precursors of these plants.

Bryophytes are like algae in other respects as well: They have neither true roots nor true leaves, they lack a vascular system, and they produce no hard tissues for structural support. The absence of these structures that are important for growth on land greatly restricts the potential size of bryophytes, which, unlike algae, are terrestrial rather than aquatic.

The ferns represent the largest group of spore-bearing vascular plants. In contrast to the bryophytes, ferns have true roots, leaves, and vascular tissues, and they produce hard tissues for support.

These architectural features enable ferns to grow to the size of small trees. Although ferns are better adapted to the drying conditions of terrestrial life than bryophytes are, they still depend on water as a medium for the movement of sperm to the egg. This dependence on water during a critical stage of their life cycle restricts the ecological range of ferns to relatively moist habitats.

The most successful terrestrial plants are the seed plants. Seed plants have been able to adapt to an extraordinary range of habitats. The embryo, protected and nourished inside the seed, is able to survive in a dormant state during unfavourable growing conditions such as drought. Seed dispersal also facilitates the dissemination of the embryos away from the parent plant.

Another important feature of seed plants is their mode of fertilization. Fertilization in seed plants is brought about by wind-or insect-mediated transfer of pollen, the gamete-producing structure of the male, to the sexual structure of the female, the pistil. Pollination is independent of external water, a distinct advantage in terrestrial environments. Many seed plants produce massive amounts of woody tissues, which enable them to grow to extraordinary heights. These features of seed plants have contributed to their success and account for their wide range.

There are two categories of seed plants: gymnosperms (from the Greek for "naked seed") and angiosperms (based on the Greek for "vessel seed," or seeds contained in a vessel). Gymnosperms are the less advanced type; about 700 species of gymnosperms are known. The largest group of gymnosperms is the conifers ("cone-bearers"), which include such commercially important forest trees as pine, fir, spruce, and redwood.

Two types of cones are present: male cones, which produce pollen, and female cones, which bear ovules. The ovules are located on the surfaces of specialized structures called cone scales. After wind-mediated pollination, the sperm reaches the egg via a pollen tube, and the fertilized egg develops into an embryo. Upon maturation, the cone scales, which are appressed during early development, separate from each other, allowing the naked seeds to fall to the ground.

Angiosperms, the more advanced type of seed plant, first became abundant during the Cretaceous period, about 100 million years ago. Today, they dominate the landscape, easily outcompeting their cousins, the gymnosperms. About 250,000 species are known, but many more remain to be characterized. A typical angiosperm life cycle, that of *Zea mays* (corn). The major innovation of the angiosperms is the flower; hence they are referred to as *flowering plants.* There are other anatomical differences between angiosperms and gymnosperms, but none so crucial and far-reaching as the mode of reproduction.

Flower Structure and the Angiosperm Life Cycle

The flower consists of several leaflike structures attached to a specialized region of the stem called the receptacle. Sepals and petals are the most leaflike. Petals have the primary function of attracting insects to serve as pollinators, accounting for their often showy and brightly coloured appearance.

The stamen is the male sexual structure, and the pistil is the female sexual structure. The pistil is composed of one or more united carpels; the pistil, or in some flowers a whorl of pistils, is sometimes referred to as the gynoecium. The stamen consists of a narrow stalk called the filament and a chambered structure called the anther. The anther contains tissue that gives rise to pollen grains. The pistil consists of the stigma (the tip where pollen lands during pollination), the style (an elongated structure), and the ovary. The ovary, the hollow basal portion of the pistil, completely encloses one or more ovules. Each ovule, in turn, contains an embryo sac, the structure that gives rise to the female gamete, the egg. After landing on the stigma, the pollen grain germinates to form a long pollen tube, which penetrates the tissues of the style and ultimately enters the cavity of the ovary, which houses the ovule. Within the ovary, the pollen tube enters the ovule and deposits two haploid sperm cells in the embryo sac. One sperm cell fuses with the egg to produce the zygote; the other typically fuses with the two polar nuclei to produce a specialized storage tissue termed the endosperm, which provides nutrients to the growing embryo.

Endosperm tissue also provides the bulk of the worldrquotes food supply in the form of cereal grains. As in conifers, in angiosperms the outer tissues of the ovule harden into a protective seed coat. Angiosperm seeds have a second layer of protective tissues, the fruit. The fruit consists of the ovary wall and, in some cases, receptacle tissue.

Angiosperms are divided into two major groups, dicotyledons (dicots) and monocotyledons (monocots). This distinction is based primarily on the number of cotyledons, or seed leaves. In addition, the two groups differ with respect to other anatomical features, such as the arrangement of their vascular tissues, and their floral structure.

As the dominant plant group on Earth, and because of their great economic and agricultural importance, angiosperms have been studied much more intensively than other types of plants. Plant physiologists have focused on a relatively small number of species that represent convenient experimental systems for the study of specific phenomena.

Therefore, while we focus on these famous few, it is important to keep in mind the tremendous diversity of form and function that exists within the angiosperms, and the even greater diversity of form and function that is found within the plant kingdom as a whole.

Plant Tissue Systems: Dermal, Ground, and Vascular

Dermal tissue. The epidermis is the dermal tissue of young plants undergoing primary growth. It is generally composed of specialized, flattened polygonal cells that occur on all plant surfaces. Shoot surfaces are usually coated with a waxy cuticle to prevent water loss and are often covered with hairs, or trichomes, which are epidermal cell extensions.

Pairs of specialized epidermal cells, the guard cells, are found surrounding microscopic pores in all leaves. The guard cells and pores are called stomata (singular stoma), and they permit gas exchange (water loss, CO_2 uptake, and O_2 release or uptake) between the atmosphere and the interior of the leaf. The root epidermis is adapted for absorption of water and minerals, and its outer wall surface typically does not have a waxy cuticle. Extensions from the root epidermal cells, the root hairs, increase the surface area over which absorption can take place.

Ground tissue. Making up the bulk of the plant are cells termed the ground tissue. There are three types of ground tissue: parenchyma, collenchyma, and sclerenchyma.

- Parenchyma, the most abundant ground tissue, consists of thin-walled, metabolically active cells that carry out a variety of functions in the plant including photosynthesis and storage
- Collenchyma tissue is composed of narrow, elongated cells with thick primary walls. Collenchyma cells provide structural support to the growing plant body, particularly shoots, and their thickened walls are nonlignified so they can stretch as the organ elongates. Collenchyma cells are typically arranged in bundles or layers near the periphery of stems or leaf petioles.

- Sclerenchyma consists of two types of cells, sclereids and fibres Both have thick secondary walls and are frequently dead at maturity. Sclereids occur in a variety of shapes, ranging from roughly spherical to branched, and are widely distributed throughout the plant. In contrast, fibres are narrow, elongated cells that are commonly associated with vascular tissues. The main function of sclerenchyma is to provide mechanical support particularly to parts of the plant that are no longer elongating.

In the stem, the pith and the cortex make up the ground tissue. The pith is located within the cylinder of vascular tissue, where it often exhibits a spongy texture because of the presence of large intercellular air spaces.

If the growth of the pith fails to keep up with that of the surrounding tissues, the pith may degenerate, producing a hollow stem. In general, roots lack piths, although there are exceptions to this rule. In contrast, the cortex, which is located between the epidermis and the vascular cylinder, is present in both stems and roots. At the boundary between the ground tissue and the vascular tissue in roots, and occasionally in stems, is a specialized layer of cortex known as the endodermis. This single layer of cells originates from cortical tissue at the innermost layer of the root cortex and forms a cylinder that surrounds the central vascular tissue, or stele.

Early in root development, a narrow band composed of the waxy substance suberin is formed in the cell walls circumscribing each endodermal cell. These suberin deposits, called Casparian strips, form a barrier in the endodermal walls to the intercellular movement of water, ions, and other water-soluble solutes to the vascular cells. Leaves have two interior layers of ground tissue that are collectively known as the mesophyll.

The palisade parenchyma consists of closely spaced, columnar cells located beneath the upper epidermis. There is usually one layer of palisade parenchyma in the leaf. Palisade parenchyma cells are rich in chloroplasts and are a primary site of photosynthesis in the leaf. Below the palisade parenchyma are i rregularly shaped, widely spaced spongy mesophyll cells. The spongy mesophyll cells are also photosynthetic, and the large spaces between these cells allow diffusion of carbon dioxide. The spongy mesophyll also contributes to leaf flexibility in the wind, and this flexibility facilitates the movement of gases within the leaf.

Vascular tissues: xylem and phloem. The vascular tissue is composed of two major conducting systems: the xylem and the phloem. The xylem transports water and mineral ions from the root to the rest of the plant.

The phloem distributes the products of photosynthesis and a variety of other solutes throughout the plant. The tracheids and vessel elements are the conducting cells of the xylem.

Both of these cell types have elaborate secondary-wall thickenings and lose their cytoplasm at maturity; that is, they are dead when functional. Tracheids overlap each other, whereas vessel elements have open end walls and are arranged end to end to form a larger unit called a vessel. Other cell types present in the xylem include parenchyma cells, which are important for the storage of energy-rich molecules and phenolic compounds, and sclerenchyma fibres.

The sieve elements and sieve cells are responsible for sugar translocation in the phloem. The former are found in angiosperms; the latter perform the same function in gymnosperms. Like vessel elements, sieve elements are often stacked in vertical rows, forming larger units called sieve tubes, whereas sieve cells form overlapping arrays. Both types of conducting cells are living when functional, but they lack nuclei and central vacuoles and have relatively few cytoplasmic organelles.

Substances are translocated from sieve cell to sieve cell laterally through circular or oval zones containing enlarged pores, called sieve areas. In contrast, sieve tubes translocate substances through large pores in the end walls of the sieve elements, called sieve plates. Sugar movement through sieve tubes is more efficient and rapid than through sieve cells and represents a more evolutionarily advanced mechanism.

Sieve elements are associated with, and depend on, densely cytoplasmic parenchyma cells called companion cells. The analogous cells adjacent to the sieve cells of gymnosperms are called albuminous cells. Companion cells provide proteins and metabolites necessary for the functions of the sieve tube elements. In addition, the phloem frequently contains storage parenchyma and fibres that provide mechanical support.

The Structures of Chloroplast Glycosylglycerides

The structures of the three plant glycosylglycerides, monogalactosyldiacylglycerol (MGDG), digalactosyldiacylglycerol (DGDG) and sulfoquinovosyldiacylglycerol (SQDG). The two galactolipids, MGDG and DGDG, are uncharged but polar; the sulfolipid, SQDG, is negatively charged as well as being polar.

The Multiple Steps in Construction of the Cell Plate Following Mitosis

During Mitosis in plants, the daughter chromosomes separate during anaphase and migrate toward their respective poles. In contrast to animal cells, the spindle poles of plant cells is quite diffuse, so that the chromosomes move in nearly parallel fashion.

At the end of teleophase, cytokinesis begins with the formation of the cell plate. This process involves fusion of many small secretory vesicles, and attachment of the resulting structure to the plasma membrane. A diagram of the steps involved in cell plate formation.

In the first step, Golgi vesicles, some of which are interconnected via *fusion tubes,* aggregate in the spindle midzone area. This structure is called the fusion tube network (FTN). The transition from the first to the second stage of cell plate formation can be inhibited by caffeine. (B) Formation of a tubulo-vesicular network (TVN).

The contents of the vesicles, mainly pectins, represent the precursors from which the new middle lamella is assembled outside the cell. In the next stage, vesicle fusion increases, forming a tubulo-vesicular network (TVN), and the membranes become coated with either clathrin or other proteins. (C) In the third stage, the central region of the growing cell plate forms a tubular network (TN), with vesicle fusion occurring at the growing edges where the remaining microtubules are located. (D) In the final stage, the cell plate contacts and adheres to the plasma membrane of the parent cell. At the same time the tubular network expands to form a fenestrated sheet. (E) At the end of mitosis, the phragmoplast disappears, the cell enters interphase, and microtubules reappear in the cytosol near the plasma membrane, where they play a role in the deposition of cellulose microfibrils during cell wall growth. (F) Electron micrograph of a cell plate forming in a root tip of a beet, (*Beta vulgaris*) (10,000×) MT, microtubule; VE, vesicles; N, nucleus; NE, nuclear envelope; P, cell plate.

Water and Plant Cells

Calculating Capillary Rise

We learn on page 36 of the textbook that the water properties of cohesion, adhesion and surface tension, give rise to the phenomenon of capillarity, the movement of water for small distances up a capillary tube. The smaller the tube radius, the higher the capillary rise. How far the water will move may be calculated using the following formula:

$$\text{Capillary rise} = \frac{14.9\times10^{-6}\,\text{m}^2}{\text{radius}}$$

where both capillary rise and radius are expressed in meters.

For a xylem vessel with 25 μm radius, the capillary rise is about 0.6 m.This distance is much too small to be significant for water transport up tall trees.

Fibrous materials such as cell walls can act like wicks to draw water by capillarity from nearby xylem.

This capillary action ensures that cell wall surfaces that are directly exposed to the air, such as those in leaf mesophyll, remain wetted and do not dry out. Because the cell wall capillaries have a tiny radius, about 10 m^{-8}, very large physical forces can be generated in the water just below the evaporative surfaces of cell walls.

Calculating Half-Times of Diffusion

Extremely Slow over Long Distances

From Fick's law, one can derive an expression for the time it takes for a substance to diffuse a particular distance.

If one defines conditions such that all the solute molecules are concentrated at the starting position, then the concentration front moves away from the starting position, as shown for a later time point. As the substance diffuses away from the starting point, the concentration gradient becomes less steep (Δc_s decreases) and thus net movement becomes slower.

The time it takes for the substance at any given distance from the starting point to reach one-half of the concentration at the starting point (t_c = ½) is given by the following equation:

$$t_{c=1/2} = \frac{(\text{dis tan ce})^2}{D_S} K$$

where K is a constant and D_s is the diffusion coefficient. The above equation shows that the time required for a substance to diffuse a given distance increases in proportion to the *square* of that distance. Let us consider two numerical examples. First, how long it would take a small molecule to diffuse across a typical cell? The diffusion coefficient for a small molecule like glucose is about 10^{-9} m² s⁻¹, and the cell length may be 50 µm. Thus, for this example:

$$t_{c=1/2} = \frac{(50+10^{-6}\,m)^2}{10^{-9}\,m^2 s^{-1}} = 2.5s$$

This calculation shows that small molecules diffuse over cellular dimensions rapidly. What about diffusion over longer distances? Calculating the time needed for the same substance to move a distance of 1 m (e.g., the length of a corn leaf), we find:

$$t_{c=1/2} = \frac{(1m)^2}{10^{-9}\,m^2 s^{-1}} = 10^9\,s \approx 32\,\text{year}$$

a value that exceeds by orders of magnitude the life span of a corn plant, which lives only a few months. This shows that diffusion in solutions can be effective within cellular dimensions but is far too slow for mass transport over long distances. Diffusion is of great importance as a driving force for the water vapour lost from leaves, because the diffusion coefficient for a molecule in air is much greater than in aqueous solutions.

Alternative Conventions for Components of Water Potential

Students planning further study of plant water relations should note that the components of water potential defined in the text are sometimes given different names and symbols. In particular, the equation

$$\psi_w = \psi_s + \psi_p$$

(textbook Equation 3.6) is often replaced by the following equivalent equation:

$$\psi_w = \pi + P$$

In this alternative convention, *P* is the same as Ψ_p. It is the hydrostatic pressure of the solution, and may be positive, as in turgid cells, or negative, as in xylem water.

The symbol π is called osmotic pressure and is the negative of Ψ_s. That is, π has positive values, and Ψ_s has negative values. "Osmotic pressure" is the term that physical chemists, zoologists, and many others use to denote the effect of dissolved solutes on the free energy of water. Most handbooks of physics and chemistry use the term "osmotic pressure" and the symbol π.

The negative sign in front of π in the equation above accounts for the reduction in water potential (Ψ_w) by dissolved solutes. Thus $\Psi_s = -\pi$. A very interesting, if somewhat unconventional, account of the history and physical meaning of osmotic pressure is given by Hammel and Scholander (1976).

Unfortunately, some authors have mixed the conventions for Ψ_s and π, leading to unnecessary confusion about what is meant by the symbol π. Thus, π is sometimes incorrectly called osmotic potential instead of osmotic pressure, and it may be used either as a positive quantity or as a negative quantity. Regarding matric potential, it is usually designated by the symbol *t*.

Can Negative Turgor Pressures Exist in Living Cells?

One of the challenging aspects of understanding plant water relations is the remarkable range of pressures, both positive and negative, that occur within the bodies of plants.

Negative pressures (or tensions), which depend upon the cohesive strength of water coupled with the strength of lignified cell walls to resist deformation, play an important role in water transport through the xylem. Positive pressures, which depend upon the semipermeable nature of the plasma membrane and the elastic nature of primary cell walls, occur in all hydrated living plants cells but can be especially large in sieve tubes and guard cells.

Living plant cells are typically assumed to have only positive pressures. However, there appears to be no reason that negative pressures could not also occur within the cytoplasm of living plant cells. This web topic explores the existence and potential role of negative turgor pressures in plants.

When a cell loses water in air, turgor declines and solute concentrations increase. At the turgor loss point ($\psi_p = 0$), the hydrostatic pressure in the cell sap is equal to atmospheric pressure, meaning that no net force is exerted on the cell wall.

If water continues to be lost from the cell, the pressure within the cytoplasm drops below atmospheric pressure, resulting in a force imbalance that collapses the cell wall. The deformation of living cells upon desiccation is called cytorrhysis. Note that the plasma membrane is pressed against the cell wall throughout desiccation (i.e., plasmolysis does not occur) because the hydrostatic pressure in cytoplasm remains greater than the hydrostatic pressure in apoplast.

In the living cells described above, a decrease in cell water potential below the turgor loss point is balanced by an increase in solute concentration as the volume of the cell decreases.

Thus, true "negative" pressures do not develop. In contrast, xylem conduits have rigid cell walls that resist deformation, allowing them to sustain negative pressures without imploding. This raises the question of what happens when water is lost from living cells that have thick walls or are embedded in a rigid matrix of cells (e.g., xylem parenchyma). Might these cells resist deformation and thus develop negative turgor pressures?

It is important to emphasize that this is a controversial area, due in large part to the absence of direct measurements of negative turgor pressures within cells. However, before reviewing the evidence for negative turgor pressures, it is worthwhile to consider what physiological effects might result from the development of such negative pressures in living cells.

The major outcome of negative turgor is that it allows stiff-walled cells to decrease in water potential without undergoing major changes in cell volume or osmotic concentration. Because cytorrhysis might cause physical damage to the wall and/or cell membranes, while the high concentrations of solutes resulting from the reduction in cell volume might adversely affect the conformation of membranes and proteins, it is possible to imagine a physiological role or benefit for the development of negative turgor pressures in plant cells.

We now turn this around and ask if there are any downsides to the generation of negative turgor pressures in living cells? In the xylem, the primary issue associated with negative pressures is cavitation; could this also be a concern for living cells?

The primary mechanism for cavitation in plants is air seeding, reflecting the fact that the probability of the de novo formation of gas voids in water (either by homogeneous or heterogeneous nucleation) is extremely low. In air seeding, air is sucked in through the cell wall, where it then "nucleates" the transition to the vapour phase.

For air seeding to occur in a living cell experiencing negative pressures, air would have to be pulled through the very small pores of the cell wall. While one can imagine this happening, the movement of air across the cell wall would result in plasmolysis and thus the immediate release of any tension in the cytoplasm because the plasma membrane is not capable of withstanding any significant pressure.

Other costs associated with the ability of living cells to generate negative pressures include the metabolic costs of producing rigid cell walls, as well as any limitations lignified walls might place on physiological function. In addition, a strategy for avoiding cell damage due to desiccation via the generation of negative pressures might impose limitations on cell size. The strength of a cell to withstand collapse (and thus generate negative pressures) is inversely proportional to cell size.

Evidence for negative turgor pressure in plants is limited, in part reflecting the fact that few researchers have devoted much attention to this issue. Living cells with flexible (unlignified) cell walls deform relatively easily upon desiccation and thus do not appear to support negative pressures. However, measurements of the forces needed to collapse cell walls suggest that living cells with thick walls can withstand forces in the range of 1.0 MPa.

Visual examination of tissues adapted to withstand very cold temperatures also provides indirect evidence for negative turgor. For example, frozen ray parenchyma cells often do not exhibit significant deformation, despite the very strong desiccatory effects (low water potential) of extracellular ice. One can hypothesize that the existence of negative pressures within these cells acts to balance the water potential gradient across the cell membrane (i.e., between the cytoplasm and ice formed within the apoplast).

In summary, negative turgor pressure remains an intriguing but little-studied area of plant water relations. While it does not appear to form in cells such as the leaf mesophyll, that can deform easily, its existence in living cells with either lignified cell walls or where the cell is embedded in a matrix of lignified tissues (e.g., living cells within wood) cannot be ruled out. The potential benefits of negative turgor pressures in terms of preventing mechanical and osmotic damage associated with severe desiccation makes.

The Matric Potential

In discussions of soils, seeds, and cell walls, one often finds reference to yet another component of Ψ_w: the matric potential (Ψ_m). The matric potential is used to account for the reduction in free energy of water when it exists as a thin surface layer, one or two molecules thick, adsorbed onto the surface of relatively dry soil particles, cell walls, and other materials.

The matric potential does not represent a new force acting on water, because the effect of surface interactions can theoretically be accounted for by an effect on Ψ_s and Ψ_p. In dry materials, however, this surface interaction effect often cannot easily be separated into Ψ_p and Ψ_s components in dry materials, so they are frequently bulked together and designated as the matric potential.

It is generally not valid to add Ψ_m to independent measurements of Ψ_s and Ψ_p to arrive at a total water potential. This is particularly true for water inside hydrated cells and cell walls, where matric effects are either negligible or they are accounted for by a reduction in Ψ_p. For instance, the negative pressure in water held by cell wall microcapillaries at the evaporative surfaces of leaves, is sometimes described as a wall matric potential. Care is needed to avoid inconsistencies when accounting for this physical effect in definitions of Ψ_p, Ψ_s, and Ψ_m.

Measuring Water Potential

Plant scientists have expended considerable effort in devising accurate and reliable methods for evaluating the water status of a plant. Four instruments that have been used extensively to measure Ψ_w, Ψ_s, and Ψ_p are described here: psychrometer, pressure chamber, cryoscopic osmometer, and pressure probe.

Psychrometer (Ψ_w Measurement)

Psychrometry (the prefix "psychro-" comes from the Greek word *psychein*, "to cool") is based on the fact that the vapour pressure of water is lowered as its water potential is reduced. Psychrometers measure the water vapour pressure of a solution or plant sample, on the basis of the principle that evaporation of water from a surface cools the surface.

Investigators make a measurement by placing a piece of tissue sealed inside a small chamber that contains a temperature sensor (in this case, a thermocouple) in contact with a small droplet of a standard solution of known solute concentration (known Ψ_s and thus known Ψ_w).

If the tissue has a lower water potential than that of the droplet, water evaporates from the droplet, diffuses through the air, and is absorbed by the tissue. This slight evaporation of water cools the drop.

The larger the difference in water potential between the tissue and the droplet, the higher the rate of water transfer and hence the cooler the droplet. If the standard solution has a lower water potential than that of the sample to be measured, water will diffuse from the tissue to the droplet, causing warming of the droplet.

Measuring the change in temperature of the droplet for several solutions of known Ψ_w makes it possible to calculate the water potential of a solution for which the net movement of water between the droplet and the tissue would be zero signifying that the droplet and the tissue have the same water potential.

Psychrometers can be used to measure the water potentials of both excised and intact plant tissue. Moreover, the method can be used to measure the Ψ_s of solutions.

This can be particularly useful with plant tissues. For example, the Ψ_w of a tissue is measured with a psychrometer, and then the tissue is crushed and the Ψ_s value of the expressed cell sap is measured with the same instrument. By combining the two measurements, researchers can estimate the turgor pressure that existed in the cells before the tissue was crushed ($\Psi_p = \Psi_w - \Psi_s$).

A major difficulty with this approach is the extreme sensitivity of the measurement to temperature fluctuations. For example, a change in temperature of 0.01°C corresponds to a change in water potential of about 0.1 MPa. Thus, psychrometers must be operated under constant temperature conditions. For this reason, the method is used primarily in laboratory settings. There are many variations in psychrometric technique; interested readers should consult Brown and Van Haveren 1972, Slavik 1974, and Boyer 1995.

Pressure Chamber

A relatively quick method for estimating the water potential of large pieces of tissues, such as leaves and small shoots, is by use of the pressure chamber. This method was pioneered by Henry Dixon at Trinity College, Dublin, at the beginning of the twentieth century, but it did not come into widespread use until P. Scholander and coworkers at the Scripps Institution of Oceanography improved the instrument design and showed its practical use. In this technique, the organ to be measured is excised from the plant and is partly sealed in a pressure chamber.

Before excision, the water column in the xylem is under tension. When the water column is broken by excision of the organ (i.e., its tension is relieved allowing its Ψ_p to rise to zero), water is pulled rapidly from the xylem into the surrounding living cells by osmosis.

The cut surface consequently appears dull and dry. To make a measurement, the investigator pressurizes the chamber with compressed gas until the distribution of water between the living cells and the xylem conduits is returned to its initial, pre-excision, state.

This can be detected visually by observing when the water returns to the open ends of the xylem conduits that can be seen in the cut surface. The pressure needed to bring the water back to its initial distribution is called the *balance pressure* and is readily detected by the change in the appearance of the cut surface, which becomes wet and shiny when this pressure is attained. The pressure chamber is often described as a tool to measure the tension in the xylem.

However, this is only strictly true for measurements made on a non-transpiring leaf or shoot (for example, one that has been previously enclosed in a plastic bag). When there is no transpiration, the water potential of the leaf cells and the water potential in the xylem will come into equilibrium.

The balancing pressure measured on such a non-transpiring shoot is equal in magnitude but opposite in sign to the pressure in the xylem (Ψ_p). Because the water potential of our non-transpiring leaf is equal to the water potential of the xylem, one can calculate the water potential of the leaf by adding together Ψ_p and Ψ_s of the xylem, provided one collects a sample of xylem sap for determination of Ψ_s.

Luckily Ψ_s of the xylem is usually small (> -0.1 MPa) compared to typical midday tensions in the xylem (Ψ_p of –1 to –2 MPa). Thus, correction for the Ψ_s of the xylem sap is frequently omitted.

Balancing pressure measurements of transpiring leaves are more difficult to interpret. The fact that water is flowing from the xylem to the leaf means that differences in water potential must exist.

When the transpiring leaf or shoot is cut off, the tension in the xylem is instantly relieved and water is drawn into the leaf cells until the water potentials of the xylem and the leaf cells come into equilibrium. Because the total volume of the leaf cells is much larger than the volume of sap in the xylem, this equilibrium water potential will be heavily weighted towards that of the leaf.

Thus, any measurement of the balancing pressure on such a leaf or shoot will result in a value that is approximately the water potential of the leaf, rather than the tension of the xylem. (To be exact, one would have to add the Ψ_s of the xylem sap to the negative of the balancing pressure to get the leaf water potential.) One can explore the differences between the water potential of the xylem and the water potential of a transpiring leaf by comparing balancing pressures measured on covered (i.e., non-transpiring) versus uncovered (transpiring) leaves.

Pressure chamber measurements provide a quick and accurate way of measuring leaf water potential. Because the pressure chamber method does not require delicate instrumentation or temperature control, it has been used extensively under field conditions. For a more complete description of the theory and operation of the pressure chamber.

Cryoscopic Osmometer

The cryoscopic osmometer measures the osmotic potential of a solution by measuring its freezing point.

Solutions have *colligative* properties that collectively depend on the number of dissolved particles and not on the nature of the solute. For example, solutes reduce the vapour pressure of a solution, raise its boiling point, and lower its freezing point.

The specific nature of the solute does not matter. One of the colligative *properties* of solutions is the decrease in the freezing point as the solute concentration increases. For example, a solution containing 1 mol of solutes per kilogram of water has a freezing point of –1.86°C, compared with 0°C for pure water. Various instruments can be used to measure the freezing-point depression of solutions. With a cryoscopic osmometer, solution samples as small as 1 nanoliter (10^{-9} L) are placed in an oil medium located on the temperature-controlled stage of a microscope. The very small sample size allows sap from single cells to be measured and permits rapid thermal equilibration with the stage.

To prevent evaporation, the investigator suspends the samples in oil-filled wells in a silver plate (silver has high thermal conductivity).

The temperature of the stage is rapidly decreased to about –30° C, which causes the sample to freeze. The temperature is then raised very slowly, and the melting process in the sample is observed through the microscope. When the last ice crystal in the sample melts, the temperature of the stage is recorded (note that the melting and freezing points are the same).

It is straightforward to calculate the solute concentration from the freezing-point depression; and from the solute concentration (c_s), Ψ_s is calculated as $-RTc_s$. This technique has been used to measure droplets extracted from single cells.

Pressure Probe

If a cell were as large as a watermelon or even a grape, measuring its hydrostatic pressure would be a relatively easy task. Because of the small size of plant cells, however, the development of methods for direct measurement of turgor pressure has been slow.

Using a micromanometer, Paul Green at the University of Pennsylvania developed one of the first direct methods for measuring turgor pressure in plant cells (Green and Stanton 1967). In this technique, an air-filled glass tube sealed at one end is inserted into a cell.

The high pressure in the cell compresses the trapped gas, and from the change in volume one can readily calculate the pressure of the cell from the ideal gas law (pressure × volume = constant). This method works only for cells of relatively large volume, such as the giant cell of the filamentous green alga *Nitella.* For smaller cells, the loss of cell sap into the glass tube is sufficient to deflate the cell and this yields artifactually low pressures.

For higher plant cells, which are several orders of magnitude smaller in volume than *Nitella,* a more sophisticated device, the pressure probe, was developed by Ernest Steudle, Ulrich Zimmermann, and their colleagues in Germany. This instrument is similar to a miniature syringe.

A glass microcapillary tube is pulled to a fine point and is inserted into a cell. The microcapillary is filled with silicone oil, a relatively incompressible fluid that can be readily distinguished from cell sap under a microscope. When the tip of the microcapillary is first inserted into the cell, cell sap begins to flow into the capillary because of the initial low pressure of that region. Investigators can observe such movement of sap under the microscope and counteract it by pushing on the plunger of the device, thus building up a pressure. In such fashion the boundary between the oil and the cell sap can be pushed back to the tip of the microcapillary.

When the boundary is returned to the tip and is held in a constant position, the initial volume of the cell is restored and the pressure inside the cell is exactly balanced by the pressure in the capillary. This pressure is measured by a pressure sensor in the device. Thus the hydrostatic pressure of individual cells may be measured directly.

This method has been used to measure Ψ_p and other parameters of water relations in cells of both excised and intact tissues of a variety of plant species. The primary limitation of this method is that some cells are too small to measure.

Furthermore, some cells tend to leak after being stabbed with the capillary, and others plug up the tip of the capillary, thereby preventing valid measurements. The pressure probe has also been adapted to measure positive and negative values of Ψ_p in the xylem. However, technical problems with cavitation limit the measurement of negative Ψ_p by this technique.

Understanding Hydraulic Conductivity

Consider a cell with an initial water potential of –0.2 MPa, submerged in pure water. From this information we know that water will flow into the cell and that the driving force is $\Delta\Psi_w = 0.2$ MPa, but what is the initial rate of movement? The rate depends on permeability of the membrane to water, a property usually called the hydraulic conductivity (*Lp*) of the membrane.

Driving force, membrane permeability, and flow rate are related by the following equation:

Flow rate = driving force × hydraulic conductivity

Hydraulic conductivity expresses how readily water can move across a membrane and has units of volume of water per unit area of membrane per unit time per unit driving force (for instance, $m^3\ m^{-2}\ s^{-1}\ MPa^{-1}$ or $m\ s^{-1}\ MPa^{-1}$). The larger the hydraulic conductivity, the larger the flow rate. The hydraulic conductivity of the membrane is $10^{-6}\ m\ s^{-1}\ MPa^{-1}$. The transport (flow) rate (J_v) can then be calculated from the following equation:

$$J_v = (Lp(\Delta\psi_w)$$

where J_v is the volume of water crossing the membrane per unit area of membrane and per unit time ($m^3\ m^{-2}\ s^{-1}$ or, equivalently, $m\ s^{-1}$). Please note that this equation assumes that the membrane is ideal—that is, that solute transport is negligible and water transport is equally sensitive to $\Delta\Psi_s$ and $\Delta\Psi_p$ across the membrane. Nonideal membranes require a more complicated equation that separately accounts for water flow induced by $\Delta\Psi_s$ and by $\Delta\Psi_p$.

In our example, J_v has a value of $0.2 \times 10^{-6}\ m\ s^{-1}$. Note that J_v has the physical meaning of a velocity. We can calculate the flow rate in volumetric terms ($m^3\ s^{-1}$) by multiplying J_v by the surface area of the cell.

The resulting value is the *initial* rate of water transport. As water is taken up, cell Ψ_w increases and the driving force ($\Delta\Psi_w$) decreases. As a result, water transport slows with time.

Wilting and Plasmolysis

Plasmolysis is the separation of plant cell cytoplasm from the cell wall as a result of water loss. It is unlikely to occur in nature, except in severe conditions. Plasmolysis is induced in the laboratory by immersing a plant cell in a strongly saline or sugary solution, so that water is lost by osmosis.

If onion epidermal tissue is immersed in a solution of calcium nitrate, cells rapidly lose water by osmosis and the protoplasm of the cells shrinks. This occurs because the calcium and nitrate ions freely permeate the cell wall and encounter the selectively permeable plasma membrane.

The large vacuole in the centre of the cell originally contains a dilute solution with much lower osmotic pressure than that of the calcium nitrate solution on the other side of the membrane. The vacuole thus loses water and becomes smaller. The space between the cell membrane and the cell wall enlarges and the plasma membrane and the protoplasm within it contract to the centre of the cell. Strands of cytoplasm extend to the cell wall because of plasma membrane-cell wall attachment points. Plasmolysed cells die unless they are transferred quickly from the salt or sugar solution to water.

Mineral Nutrition

Symptoms of Deficiency in Essential Minerals

Visual nutrient deficiency symptoms can be a very powerful diagnostic tool for evaluating the nutrient status of plants. One should keep in mind, however, that a given individual visual symptom is seldom sufficient to make a definitive diagnosis of a plant's nutrient status.

Many of the classic deficiency symptoms such as tip burn, chlorosis and necrosis are characteristically associated with more than one mineral deficiency and also with other stresses that by themselves are not diagnostic for any specific nutrient stress. However, their detection is extremely useful in making an evaluation of nutrient status.

In addition to the actual observations of morphological and spectral symptoms, knowing the location and timing of these symptoms is a critical aspect of any nutrient status evaluation. Plants do not grow in isolation, they are part of the overall environment and as such they respond to environmental changes as that affect nutrient availability. Also, plants do influence their environment and can contribute to environmental changes, which in turn can affect the nutrient status of the plant.

Sources of Visual Symptoms

Stresses such as salinity, pathogens, and air pollution induce their own characteristic set of visual symptoms. Often, these symptoms closely resemble those of nutrient deficiency.

Pathogens often produce an interveinal chlorosis, and air pollution and salinity stress can cause tip burn. Although at first these symptoms might seem similar in their general appearance to nutrient deficiency symptoms, they do differ in detail and/or in their overall developmental pattern.

Pathological symptoms can often be separated from nutritional symptoms by their distribution in a population of affected plants. If the plants are under a nutrient stress, all plants of a given type and age in the same environment tend to develop similar symptoms at the same time.

However if the stress is the result of pathology, the development of symptoms will have a tendency to vary between plants until a relatively advanced stage of the pathology is reached.

Environmental Associations

Plants remove substantial amounts of nutrients from the soil during their normal growth cycle and many long-term environmental changes occur as a result of this process. Effects on the soil go considerably beyond the straight removal or depletion of nutrients. Charge balance must be maintained in the plant-soil system during nutrient uptake. Charge balance is usually achieved by the excretion of proton and/or hydroxyl ions by the plant to replace the absorbed nutrient cations or anions.

For example when plants are fertilized with ammonia, they acquire most of their nitrogen in the form of the ammonium cation, rather than from the usual nitrate anion. Because nitrate is the only anion used by the plant in large amounts, the net result of this change is that during normal nutrient uptake the proton excretion will far exceed that of hydroxyl ions.

In the case of vigorously growing plants, the amount of excreted protons can be sufficiently large as to decrease the pH of the soil by several pH units. Changes in soil pH of such magnitude can have large implications for a number of soil processes such as soil structure, nutrient availability and leaching of nutrients.

The immediate effect on the soil may be favourable for some plants, especially acid-loving plants, in that it tends to make iron more available. However, in the long run, lowering the soil pH can be deleterious to plants in that the availability of nutrients will change. A lower soil pH will allow micronutrients to be more readily leached from the soil profile, eventually resulting in deficiencies of nutrients such as Cu and Zn. Additionally, when the pH of the soil drops much below pH 5, the solubility of Al and Mn can increase to such an extent as to become toxic to most plant growth.

Plants are often thought of as passive in relation to the environment. However this is not always a valid assumption; for there are many plants that clearly manipulate their environment in a fashion that tends to makes certain nutrients more readily available.

For example, iron is a limiting nutrient in many agricultural areas, but it comprises about 3% of the average soil which, if available, would be far in excess of the needs of the average plant. Some plants actively excrete protons, and the resulting decrease in pH increases the solubility of iron in their environment. In addition, other plants excrete phytosiderophores that chelate the soil iron rendering it a more available form for the plants.

Pathways of Symptom Development

At first glance, it would appear that the distinction of deficiency symptoms for the 13 known essential mineral nutrients should be relatively simple. But such an assumption is incorrect. In fact, the deficiency symptoms are quite complex because each nutrient has a number of different biological functions and each function may have an independent set of interactions with a wide range of environmental parameters.

In addition, the expression of these symptoms varies for acute or chronic deficiency conditions. Acute deficiency occurs when a nutrient is suddenly no longer available to a rapidly growing plant. Chronic deficiency occurs when there is a limited but continuous supply of a nutrient, at a rate that is insufficient to meet the growth demands of the plant. Most of the classic deficiency symptoms described in textbooks are characteristic of acute deficiencies. The most common symptoms of low-grade, chronic deficiencies are a tendency towards darker green leaves and stunted or slow growth.

Typically most published descriptions of deficiency symptoms arise from experiments conducted in greenhouses or growth chambers where the plants are grown in hydroponics or in media where the nutrients are fully available. In these conditions, nutrients are readily available while present, but when a nutrient is depleted, the plant suddenly faces an acute deficiency. Thus, hydroponic studies favour the development of acute deficiencies.

In experiments designed to study micronutrient deficiency symptoms, micronutrients are usually omitted from the nutrient solution. Micronutrients are often present in the seed or as contaminants in the environment, so a plant of adequate size will exhaust these trace amounts of micronutrient and develop characteristic acute deficiency systems.

When deficiency symptoms of macronutrients are sought, the macronutrient is removed suddenly from a suitable sized rapidly growing plant. Alternatively the plant can initially be given a one-time supply of the nutrient that is sufficient for a limited amount of growth. Because macronutrients are continuously required in relatively large amounts by rapidly growing plants, the available nutrients will be rapidly depleted, resulting in an acute deficiency.

In natural systems, the plant encounters many degrees and types of stresses that result in different types of symptoms occurring over time. Perhaps the most common nutrient deficiency in natural environments is the case of a limited nutrient supply that is continuously renewed at a low rate from soil weathering processes. In such cases, the limited nutrient availability results in chronic nutrient deficiency symptoms.

Effect of Nutrient Mobility on Symptom Development

The interaction between nutrient mobility in the plant, and plant growth rate can be a major factor influencing the type and location of deficiency symptoms that develop. For very mobile nutrients such as nitrogen and potassium, deficiency symptoms develop predominantly in the older and mature leaves. This is a result of these nutrients being preferentially mobilized during times of nutrient stress from the older leaves to the newer leaves near the growing regions of the plant. Additionally, mobile nutrients newly acquired by the roots are also preferentially translocated to new leaves and the growing regions.

Thus old and mature leaves are depleted of mobile nutrients during times of stress while the new leaves are maintained at a more favourable nutrient status.

The typical localization of deficiency symptoms of very weakly mobile nutrients such as calcium, boron, and iron is the opposite to that of the mobile nutrients; these deficiency symptoms are first displayed in the growing regions and new leaves while the old leaves remain in a favourable nutrient status. (This assumes that these plants started with sufficient nutrient, but ran out of nutrient as they developed).

In plants growing very slowly for reasons other than nutrition (such as low light) a normally limiting supply of a nutrient could, under these conditions, be sufficient for the plant to slowly develop, maybe even without symptoms. This type of development is likely to occur in the case of weakly mobile nutrients because excess nutrients in the older leaves will eventually be mobilized to supply newly developing tissues.

In contrast, a plant with a similar supply that is growing rapidly will develop severe deficiencies in the actively growing tissue such as leaf edges and the growing region of the plant. A classic example of this is calcium deficiency in vegetables such as lettuce where symptoms develop on the leaf margins (tip burn) and the growing region near the meristems.

The maximal growth rate of lettuce is often limited by the internal translocation rate of calcium to the growing tissue rather than from a limited nutrient supply in the soil.

When moderately mobile nutrients such as sulphur and magnesium are the limiting nutrients of the system, deficiency symptoms are normally seen over the entire plant. However the growth rate and rate of nutrient availability can make a considerable difference on the locations at which the symptoms develop. If the nutrient supply is marginal compared to the growth rate, symptoms will appear on the older tissue, but if the nutrient supply is very low compared to the growth rate, or the nutrient is totally depleted, the younger tissue will become deficient first.

Plant Competition and Induced Deficiencies

When the observed symptoms are the direct result of a nutrient deficiency, the actions needed for correction are relatively straight-forward. However symptoms are often the result of interactions with other environmental factors limiting the availability of the nutrient whose symptoms are expressed. The classic instance is that of iron deficiency induced by an excess of heavy metals in the environment. Transition metals such as Cu, Zn Cr and Ni compete with Fe and each other for plant uptake. Competition for uptake is not specific to Fe and heavy metals but is true for all mineral nutrients that are chemically similar and have similar uptake mechanisms. For example if the availability of Cu or Zn is relatively less than that of Fe, then excessive concentrations of some other metal such as Ni or Cr will induce a deficiency of one of these nutrients rather than Fe. In the case of the macronutrients, excessive amounts of Mg will compete with K for uptake and can possibly induce a K deficiency. The barrenness of serpentine soils is the result of such competition, with the high Mg of these soils inducing a Ca deficiency. The toxicity of a low pH soil is another example of a basic nutrient deficiency. Low pH has a two-fold effect on soil nutrients: It enhances the leaching of cations, reducing their availability in the soil, and the relatively abundant protons in the soil compete with Ca and other cations for uptake. Thus, nutrient deficiencies can be induced by a number of different mechanisms often working in concert to limit the availability of a nutrient.

Nutrient Demand and Use Efficiency

Although all plants of the same species respond similarly to nutrient stress, plants of similar species will often show significant differences in their nutrient use efficiency. This results from differences in growth rate, root distribution, phase of development, and efficiency of nutrient uptake and utilization. This implies that in any given location, plants from one species may become nutrient-deficient, while those from another species growing in the same environment right next to them, may not show any deficiency symptoms.

Growth rate also affects nutrient status. When the nutrient supply is barely inadequate for growth under existing environmental conditions, many plants adjust their growth rate to match that supported by the available nutrient supply without displaying typical visual deficiency symptoms. Agricultural systems differ from natural systems in that crop plants have been selected primarily for rapid growth under low stress conditions.

This rapid growth rate results in a high nutrient demand by these plants and a higher incidence of nutrient deficiency unless supplemental fertilizers are supplied. It is not uncommon to find agricultural crops showing severe signs of nutrient stress, with native plants growing in the same area showing little or no indication of nutrient stress. In agriculture systems chronic deficiency symptoms develop mostly in crops with little or limited fertilization. Acute nutrient deficiency symptoms most often occur when new crops with a higher nutrient demand are introduced, or less productive lands are brought under cultivation for the production of rapidly growing crop plants.

Uniformity of Nutrient Status

Not all tissues of a plant are at the same nutrient status during times of stress. Leaves on the same plant that are exposed to different environmental conditions, (such as light), or those of different ages may have considerable differences in nutrient status. Mineral nutrients are for the most part acquired by the roots and translocated throughout the plant. The distance of any part of the plant to the roots will influence nutrient availability, particularly in the case of the less mobile nutrients. In plants recovering from nutrient deficiency, the root and conductive tissues recover first. For example, in the case of recovery from Fe deficiency, it is common to see the veins re-green while the interveinal tissue remains chlorotic and Fe-deficient.

In order to maintain rapid, optimal growth, all plant tissues must have a favourable nutrient status. Although a plant may be marginally low in a number of nutrients, only one nutrient at a time will limit overall growth. However, if the supply of that limiting nutrient is increased even slightly, the resulting increase in growth will increase the demand for all other nutrients and another nutrient, the next lowest in availability, will become limiting.

Other Diagnostic Tools

Although visual diagnostic symptoms are an extremely valuable tool for the rapid evaluation of the nutrient status of a plant, they are only some of the tools available. Other major tools include microscopic studies, spectral analysis, and tissue and soil analysis. These methods all vary in their precision, rapidity and their ability to predict future nutrient status. Because of the close interaction between plant growth and the environment, all predictions of future nutrient status must make assumptions about how the environment will change in that time frame.

The principle advantage of visual diagnostic symptoms is that they are readily obtained and provide an immediate evaluation of nutrient status. Their main drawback is that the visual symptoms do not develop until after there has been a major effect on yield, growth and development.

Tissue analysis is nutrient-specific but relatively slow; tissues must be sampled, processed and analysed before the nutrient status can be determined. An analysis of the mineral nutrient content of selected plants tissues, when compared against Critical Level values, can be used to evaluate the plant nutrient status at the time of sampling with a relatively high degree of confidence and can be extrapolated to project nutrient status at harvest. Soil analysis is similar to tissue analysis but evaluates the potential supplying power of the soil instead of plant nutrient status. Plant analysis provides information as to what the plant needs, while soil analysis provides information about the status of the nutrient supply.

Spectral analysis of nutrient status is still in its infancy and is presently used primarily in the inventory of global resources and in specialized studies. Microscopic studies are most valuable in looking at the physiological aspects of nutrient stress rather than the evaluation of plant nutrient status on a whole plant or crop basis.

Symptom Descriptions

It is unusual to find any one leaf or even one plant that displays the full array of symptoms that are characteristic of a given deficiency. It is thus highly desirable to know how individual symptoms look, for it is possible for them to occur in many possible combinations on a single plant. Most of the terms used below in the description of deficiency symptoms are reasonably self evident; a few however have a distinct meaning in the nutrient deficiency field. For example, the term *chlorotic,* which is a general term for yellowing of leaves through the loss of chlorophyll, cannot be used without further qualification because there may be an overall chlorosis as in nitrogen deficiency, interveinal, as in iron deficiency, or marginal, as in calcium deficiency. Another term used frequently in the description of deficiency symptoms is *necrotic,* a general term for brown, dead tissue. This symptom can also appear in many varied forms, as is the case with chlorotic symptoms.

Nutrient deficiency symptoms for many plants are similar, but because of the large diversity found in plants and their environments there is a range of expression of symptoms. Because of their parallel veins, grasses and other monocots generally display the affects of chlorosis as a series of stripes rather than the netted interveinal chlorosis commonly found in dicots. The other major difference is that the marginal necrosis or chlorosis found in dicots is often expressed as tip burn in monocots.

Magnesium. The Mg-deficient leaves show advanced interveinal chlorosis, with necrosis developing in the highly chlorotic tissue. In its advanced form, magnesium deficiency may superficially resemble potassium deficiency. In the case of magnesium deficiency the symptoms generally start with mottled chlorotic areas developing in the interveinal tissue. The interveinal laminae tissue tends to expand proportionately more than the other leaf tissues, producing a raised puckered surface, with the top of the puckers progressively going from chlorotic to necrotic tissue. In some plants such as the Brassica (The mustard family, which includes vegetables such as broccoli, brussel sprouts, cabbage, cauliflower, collards, kale, kohlrabi, mustard, rape, rutabaga and turnip.), tints of orange, yellow, and purple may also develop.

Manganese. These leaves show a light interveinal chlorosis developed under a limited supply of Mn. The early stages of the chlorosis induced by manganese deficiency are somewhat similar to iron deficiency. They begin with a light chlorosis of the young leaves and netted veins of the mature leaves especially when they are viewed through transmitted light. As the stress increases, the leaves take on a gray metallic sheen and develop dark freckled and necrotic areas along the veins. A purplish luster may also develop on the upper surface of the leaves. Grains such as oats, wheat, and barley are extremely susceptible to manganese deficiency. They develop a light chlorosis along with gray specks which elongate and coalesce, and eventually the entire leaf withers and dies.

Molybdenum. These leaves show some mottled spotting along with some interveinal chlorosis. An early symptom for molybdenum deficiency is a general overall chlorosis, similar to the symptom for nitrogen deficiency but generally without the reddish coloration on the undersides of the leaves. This results from the requirement for molybdenum in the reduction of nitrate, which needs to be reduced prior to its assimilation by the plant.

Thus, the initial symptoms of molybdenum deficiency are in fact those of nitrogen deficiency. However, molybdenum has other metabolic functions within the plant, and hence there are deficiency symptoms even when reduced nitrogen is available. In the case of cauliflower, the lamina of the new leaves fail to develop, resulting in a characteristic whiptail appearance. In many plants there is an upward cupping of the leaves and mottled spots developing into large interveinal chlorotic areas under severe deficiency.

At high concentrations, molybdenum has a very distinctive toxicity symptom in that the leaves turn a very brilliant orange.

Nitrogen. The chlorotic symptoms shown by this leaf resulted from nitrogen deficiency. A light red cast can also be seen on the veins and petioles. Under nitrogen deficiency, the older mature leaves gradually change from their normal characteristic green appearance to a much paler green. As the deficiency progresses these older leaves become uniformly yellow (chlorotic). Leaves approach a yellowish white colour under extreme deficiency. The young leaves at the top of the plant maintain a green but paler colour and tend to become smaller in size. Branching is reduced in nitrogen deficient plants resulting in short, spindly plants. The yellowing in nitrogen deficiency is uniform over the entire leaf including the veins. However in some instances, an interveinal necrosis replaces the chlorosis commonly found in many plants. In some plants the underside of the leaves and/or the petioles and midribs develop traces of a reddish or purple colour. In some plants this coloration can be quite bright. As the deficiency progresses, the older leaves also show more of a tendency to wilt under mild water stress and become senescent much earlier than usual. Recovery of deficient plants to applied nitrogen is immediate (days) and spectacular.

Phosphorus. These phosphorus-deficient leaves show some necrotic spots. As a rule, phosphorus deficiency symptoms are not very distinct and thus difficult to identify. A major visual symptom is that the plants are dwarfed or stunted. Phosphorus deficient plants develop very slowly in relation to other plants growing under similar environmental conditions but without phosphorus deficiency. Phosphorus deficient plants are often mistaken for unstressed but much younger plants. Some species such as tomato, lettuce, corn and the brassicas develop a distinct purpling of the stem, petiole and the under sides of the leaves. Under severe deficiency conditions there is also a tendency for leaves to develop a blue-gray luster. In older leaves under very severe deficiency conditions a brown netted veining of the leaves may develop.

Sulphur. This leaf shows a general overall chlorosis while still retaining some green colour. The veins and petioles show a very distinct reddish colour. The visual symptoms of sulphur deficiency are very similar to the chlorosis found in nitrogen deficiency. However, in sulphur deficiency the yellowing is much more uniform over the entire plant including young leaves. The reddish colour often found on the underside of the leaves and the petioles has a more pinkish tone and is much less vivid than that found in nitrogen deficiency. With advanced sulphur deficiency brown lesions and/or necrotic spots often develop along the petiole, and the leaves tend to become more erect and often twisted and brittle.

Zinc. This leaf shows an advanced case of interveinal necrosis. In the early stages of zinc deficiency the younger leaves become yellow and pitting develops in the interveinal upper surfaces of the mature leaves. Guttation is also prevalent. As the deficiency progress these symptoms develop into an intense interveinal necrosis but the main veins remain green, as in the symptoms of recovering iron deficiency. In many plants, especially trees, the leaves become very small and the internodes shorten, producing a rosette like appearance.

Boron. These boron-deficient leaves show a light general chlorosis. The tolerance of plants to boron varies greatly, to the extent that the boron concentrations necessary for the growth of plants having a high boron requirement may be toxic to plants sensitive to boron. Boron is poorly transported in the phloem of most plants, with the exception of those plants that utilize complex sugars, such as sorbitol, as transport metabolites. In a recent study, tobacco plants engineered to synthesize sorbitol were shown to have increased boron mobility, and to better tolerate boron deficiency in the soil.

In plants with poor boron mobility, boron deficiency results in necrosis of meristematic tissues in the growing region, leading to loss of apical dominance and the development of a rosette condition. These deficiency symptoms are similar to those caused by calcium deficiency. In plants in which boron is readily transported in the phloem, the deficiency symptoms localize in the mature tissues, similar to those of nitrogen and potassium. Both the pith and the epidermis of stems may be affected, often resulting in hollow or roughened stems along with necrotic spots on the fruit. The leaf blades develop a pronounced crinkling and there is a darkening and crackling of the petioles often with exudation of syrupy material from the leaf blade. The leaves are unusually brittle and tend to break easily. Also, there is often a wilting of the younger leaves even under an adequate water supply, pointing to a disruption of water transport caused by boron deficiency.

Calcium. These calcium-deficient leaves show necrosis around the base of the leaves. The very low mobility of calcium is a major factor determining the expression of calcium deficiency symptoms in plants. Classic symptoms of calcium deficiency include blossom-end rot of tomato (burning of the end part of tomato fruits), tip burn of lettuce, blackheart of celery and death of the growing regions in many plants. All these symptoms show soft dead necrotic tissue at rapidly growing areas, which is generally related to poor translocation of calcium to the tissue rather than a low external supply of calcium. Very slow growing plants with a deficient supply of calcium may re-translocate sufficient calcium from older leaves to maintain growth with only a marginal chlorosis of the leaves. This ultimately results in the margins of the leaves growing more slowly than the rest of the leaf, causing the leaf to cup downward. This symptom often progresses to the point where the petioles develop but the leaves do not, leaving only a dark bit of necrotic tissue at the top of each petiole. Plants under chronic calcium deficiency have a much greater tendency to wilt than non-stressed plants.

Chloride. These leaves have abnormal shapes, with distinct interveinal chlorosis. Plants require relatively high chlorine concentration in their tissues. Chlorine is very abundant in soils, and reaches high concentrations in saline areas, but it can be deficient in highly leached inland areas. The most common symptoms of chlorine deficiency are chlorosis and wilting of the young leaves. The chlorosis occurs on smooth flat depressions in the interveinal area of the leaf blade. In more advanced cases there often appears a characteristic bronzing on the upper side of the mature leaves. Plants are generally tolerant of chloride, but some species such as avocados, stone fruits, and grapevines are sensitive to chlorine and can show toxicity even at low chloride concentrations in the soil.

Copper. These copper-deficient leaves are curled, and their petioles bend downward. Copper deficiency may be expressed as a light overall chlorosis along with the permanent loss of turgor in the young leaves. Recently matured leaves show netted, green veining with areas bleaching to a whitish gray. Some leaves develop sunken necrotic spots and have a tendency to bend downward. Trees under chronic copper deficiency develop a rosette form of growth. Leaves are small and chlorotic with spotty necrosis.

Iron. These iron-deficient leaves show strong chlorosis at the base of the leaves with some green netting. The most common symptom for iron deficiency starts out as an interveinal chlorosis of the youngest leaves, evolves into an overall chlorosis, and ends as a totally bleached leaf. The bleached areas often develop necrotic spots. Up until the time the leaves become almost completely white they will recover upon application of iron. In the recovery phase the veins are the first to recover as indicated by their bright green colour. This distinct venial re-greening observed during iron recovery is probably the most recognizable symptom in all of classical plant nutrition. Because iron has a low mobility, iron deficiency symptoms appear first on the youngest leaves. Iron deficiency is strongly associated with calcareous soils and anaerobic conditions, and it is often induced by an excess of heavy metals.

Potassium. Some of these leaves show marginal necrosis (tip burn), others at a more advanced deficiency status show necrosis in the interveinal spaces between the main veins along with interveinal chlorosis. This group of symptoms is very characteristic of K deficiency symptoms. The onset of potassium deficiency is generally characterized by a marginal chlorosis progressing into a dry leathery tan scorch on recently matured leaves. This is followed by increasing interveinal scorching and/or necrosis progressing from the leaf edge to the midrib as the stress increases. As the deficiency progresses, most of the interveinal area becomes necrotic, the veins remain green and the leaves tend to curl and crinkle. In some plant such as legumes and potato, the initial symptom of deficiency is white speckling or freckling of the leaf blades. In contrast to nitrogen deficiency, chlorosis is irreversible in potassium deficiency, even if potassium is given to the plants. Because potassium is very mobile within the plant, symptoms only develop on young leaves in the case of extreme deficiency. Potassium deficiency can be greatly alleviated in the presence of sodium but the resulting sodium-rich plants are much more succulent than a high potassium plant. In some plants over 90% of the required potassium can be replaced with sodium without any reduction in growth.

Observing Roots below Ground

Studying root growth below the soil surface requires a means of observing the root system directly. As early as 1873, the German botanist Julius von Sachs studied root systems by using simple soil-filled boxes with one glass wall. Since that time, facilities for studying root growth in the soil have become much more complex. Large laboratories with subterranean chambers for the observation of root growth have been constructed and allow the analysis of root growth while the aerial parts of the plant are exposed to natural field conditions. These laboratories are called rhizotrons (from the Greek *rhizos*, meaning "root," and *tron*, meaning "a device for studying"). In a rhizotron, roots grow in glass-walled chambers that line underground passageways. Grid lines on the glass walls indicate soil depth. Details of root morphology (root size and distribution) under natural growing conditions can be observed with specially designed microscopes, mounted adjacent to the glass walls of the root chambers. In addition, the growth of the roots over a period of time can be measured with time-lapse photography.

Because rhizotrons are expensive to construct and maintain, they have been largely supplanted by minirhizotrons or root periscopes. Minirhizotrons are transparent plastic tubes previously buried at an angle in the soil near the plants to be observed. An optical device such as a tilted mirror with a magnifying lens or a miniature videocamera is inserted into a tube to monitor roots growing along its surface. This device provides information on rooting density in the bulk soil, as well as on root growth and phenology. Plant physiologists use such information in conjunction with changes in nutrient and water levels in the soil to assess root activity through the soil profile.

Solute Transport

Patch Clamp Studies in Plant Cells

The use of the patch clamp method in plant physiology is generating information about the properties of plant membranes that is beyond the scope of other techniques. Conventional electrophysiological studies with intracellular electrodes are valuable for the measurement of membrane potentials and other electrical properties of plant cells, but these methods have limitations for the characterization of ion pumps and single ion channels. Patch clamping, on the other hand, is very well suited for that purpose.

During conventional intracellular recording of cellular electric potentials, cells are impaled with a glass micropipette electrode, and the electrode tip is located inside the cell, usually in the vacuole. For patch clamp experiments, the cell walls are first digested away by treatment with the appropriate cell wall enzymes, resulting in naked protoplasts. The tip of a glass micropipette is then brought into contact with the plasma membrane of the protoplast, and gentle suction is applied to facilitate the formation of a tight seal between the micropipette and the membrane. The unusual tightness of the seal that forms between the patch pipette and the membrane reduces the leakage-caused background electrical noise typical of conventional intracellular recording sufficiently to allow high-resolution recordings of the currents through single ion channels.

When a tight seal is attained, several options become available to the investigator. Further suction can remove the portion of the membrane delimited by the opening in the pipette tip, exposing the interior of the cell to the micropipette solution (thus different solutions can be introduced into the cell through the micropipette). In this configuration, measured electric currents reflect the sum total of all electric charges carried by active and passive ion fluxes across the plasma membrane of the whole cell. Pulling the electrode away from the cell produces a membrane patch that becomes exposed to both the pipette solution and the bathing medium. Because the membrane patches contain only a few channels, this configuration makes it possible to study the abrupt changes in current caused by the opening and closing of single ion channels. Remarkably, this technique is sensitive enough to detect the change in conformation of a single protein molecule. Because protoplasts are spherical, their volume is easy to measure, and their lack of electrical connections with other cells facilitates calculations of charge fluxes per unit area of membrane. Other advantages of patch clamping are the ability to distinguish between electrical events at the plasma membrane and the tonoplast and the possibility of controlling the composition of both the cytosol and the bathing medium.

The gradual increase in current after a change in voltage reflects the opening of increasing numbers of channels in response to the membrane potential. Channels responding in this way are termed voltage-gated channels. In general, the gating (opening and closing) of a channel is controlled independently of its ion selectivity, which is generally constant for a particular kind of channel protein. Again the current does not depend in a simple way on the driving force for K^+ diffusion (that is, the deviation of the potential from the K^+ equilibrium potential). An outward current is activated only at potentials more positive than -40 mV, and an inward current is activated only at potentials more negative than about -100 mV. This activation is a reflection of the gating of individual channel proteins with different properties.

The inward and outward currents can be shown to have different single-channel conductances (the ability of current to cross the membrane) and varying response times, and they are differentially affected by various activators and inhibitors. It is now known that the inward and outward currents are carried by two different kinds of K^+ channels, as demonstrated by the cloning and expression of the respective genes. The outward-rectifying channels open only when the membrane potential favours outward diffusion of K^+, and the inward-rectifying channels open only when the membrane potential favours the inward diffusion of K^+. Since each type of channel has its own regulatory mechanisms, the cell can independently control the rates of inward and outward diffusion. Additional information about K^+ channels can be obtained from studies of currents in isolated patches of membrane in which single channels can be visualized. When held at a constant voltage, these patches show abrupt jumps in the level of current, as single K^+ channels open and close.

This behaviour is seen only in membrane channels, never in ion pumps or carriers, because a single conformational change of a channel can allow hundreds of thousands of ions to cross the membrane, whereas a conformational change of a pump or a carrier can transport only one, or a few, molecules or ions. Patch clamp studies have also shown that gated channels can be distinguished on the basis of how long the gates remain open in response to a prolonged stimulus. Two types of anion channels have been characterized at the guard cell plasma membrane: Rapidly activated (R-type) and slowly activated (S-type). *R-type channels* open and close very rapidly in response to a voltage stimulus; *S-type channels* remain open for the duration of the stimulus. Rapid and slow channels have also been identified in vacuolar membranes, and they are referred to as *fast vacuolar (FV)* or *slow vacuolar (SV)* channels.

Kinetic Analysis of Multiple Transporter Systems

Studies of K_m values have shown a feedback inhibition of carrier-mediated transport that allows the cell to maintain stable cytoplasmic ion concentrations over a wide range of external concentrations.

For example, the apparent K_m value of the high-affinity K^+ carrier in barley roots changes with the plant's demand for K^+. Barley plants starved for K^+ have high-affinity K_m values in the range of 0.02 to 0.03 m*M*, whereas well-fertilized plants have K_m values that are four-to fivefold greater. An increase in K_m reflects a decrease in the carrier's affinity for K^+. The change in apparent $K,_m$ probably reflects the expression of different members of a gene family, with their gene products thereby matching cellular needs to environmental ionic conditions. If the plants' adjustments were merely a matter of adding more transporters to the plasma membrane, there would be change in V_{max} without a change in K_m.

This is not what is observed for K^+. However, plants that have been starved and then resupplied with sulfate or phosphate often respond by an increase in V_{max}. As the internal sulfate and phosphate levels recover, the V_{max} values for their uptake drop to the usual values as excess carriers are removed from the membrane.

ABC Transporters in Plants

The ATP-binding cassette (ABC) transporters are members of a large family of active transport proteins energized directly by ATP hydrolysis. Originally identified in microbial and animal cells, ABC transporters are able to use the energy of ATP hydrolysis directly to pump organic molecules (especially large anionic molecules) across a membrane. Like the P-type H^+-ATP synthases, ABC transporters form a phosphorylated intermediate during catalysis, and they are therefore inhibited by vanadate. A number of genes for ABC transporters have been cloned from various plants. Genes for ABC proteins are recognizable by a predicted protein structure consisting of one or two copies each of two structural units, a transmembrane domain (crossing the membrane several times) and a nucleotide-binding fold, which contains a characteristic signature sequence. The transmembrane domain provides the pathway for solute transport and determines the molecular specificity of the transporter. An ABC transporter has been found to secrete anti-fungal terpenes across the plasma membrane of tobacco cells. More commonly, ABC transporters are found at the tonoplast, where they are sometimes referred to as glutathione conjugate pumps, or GS-X pumps, since they often transport molecules that have been covalently attached to glutathione. Glutathione (GS) is a tripeptide (Glu-Cys-Gly) that functions as an important cellular antioxidant molecule. The vacuolar GS-X pumps of plant cells function in herbicide detoxification, protection against oxidative damage, pigment accumulation, and the storage of antimicrobial compounds.

A family of enzymes called glutathione transferases (GSTs) is responsible for attaching glutathione to the molecule to be transported. The common chemical determinant for GS conjugation is a carbon–carbon double bond adjacent to an electron-withdrawing group (CH_2 = CH-Z). Compounds that contain this determinant include the anthocyanins, IAA (indoleacetic acid, or auxin), and various phenolic compounds. Metal-binding polypeptides called phytochelatins, which are synthesized from glutathione, are also transported into plant vacuoles along with their chelated heavy metals via GS-X pumps.

Transport Studies with Isolated Vacuoles and Membrane Vesicles

Just as studies with isolated thylakoid membranes led to the elucidation of the electron transfer reactions of photosynthesis, the development of techniques for isolating functionally competent tonoplasts and plasma membranes has enabled scientists to study the transport reactions of these membranes in vitro. The use of isolated membrane vesicles to study transport avoids interference from metabolism. Since the contents of the vesicles can be fairly readily exchanged, the concentrations of solutes on each side of the membrane can be better defined. Two types of preparations have been used: intact vacuoles, and isolated plasma membrane or tonoplast vesicles.

Isolating Intact Vacuoles from Protoplasts

When protoplasts prepared by enzymatic digestion of the cell walls are maintained in the plasmolyzed state by incubation in medium containing a suitable osmotic solute, such as mannitol, the plasma membranes can be carefully lysed without breaking the tonoplast by a gentle osmotic shock. The intact vacuoles that are released by this treatment can then be purified by centrifugation through a sucrose density gradient. Intact vacuoles (along with contaminating cellular debris) are centrifuged in a tube containing a concentration gradient of sucrose (or another suitable solute). Because intact vacuoles tend to exclude the solute, they sediment to a lighter density on the density gradient than the other organelles and membranes do. In this way preparations of highly purified intact vacuoles can be obtained.

Isolating Tonoplast Membrane from Cell Homogenates

Alternatively, tonoplasts and plasma membranes can be purified as small membrane vesicles. When tissues are ground in a mortar and pestle or homogenized in a blender, broken pieces of various membranes tend to seal up to form small vesicles less than 1 μm in diameter. Because of a low ratio of protein to lipid in the tonoplast membrane, tonoplast vesicles usually have a relatively low density, and they can be separated from other cellular membranes by density-gradient centrifugation.

Isolating Plasma Membrane from Cell Homogenates

Plasma membrane vesicles can be purified by phase partition in a mixture of dextran and polyethylene glycol solutions. Such a mixture separates into two phases, and because of their surface properties, plasma membranes tend to accumulate in the upper phase, while intracellular membranes tend to remain in the lower phase. Tissue homogenization often produces a mixture of vesicles of different orientation (right-side-out and inside-out vesicles). In the case of plasma membrane vesicles, phase partition selects for right-side-out vesicles, which therefore show little reactivity to ATP because the catalytic sites of the enzymes are facing inside the vesicle. If a detergent is added to such a preparation, ATP penetrates the vesicles and is hydrolyzed, but since the vesicles are no longer sealed, no transport is observed. One kind of detergent in low doses does not make the membranes leaky but, it is claimed, causes them to reverse their orientation. Apparently, this detergent accumulates in the inner half of the lipid bilayer, causing it to expand and flip the vesicles inside out. Since the catalytic site of the enzyme is now facing the external solution, in the presence of this detergent the vesicles respond to ATP by creating a gradient of pH and membrane potential.

Measuring Transport in Membrane Vesicles

Radioisotopes can be used to monitor the transport of various solutes into membrane vesicles. The vesicles are incubated for various times with the labelled solute, then separated by rapid filtration, and their radioactivity is measured in a scintillation counter. Other methods have been developed to measure the gradients of pH and membrane potential created by proton pumps. One such method uses fluorescent weak bases such as 9-aminoacridine (9-AA) or quinacrine, which are lipid-soluble in their uncharged form and readily diffuse across the membrane and enter the vesicle. Within the acidified vesicle, the amino group becomes protonated, trapping the charged form of the molecule inside. As the dye molecules accumulate inside the vesicles, they form dimers and other aggregates that result in a reduction in their fluorescence, a phenomenon termed fluorescence quenching. One can thus study ATP-driven proton transport across the vesicle membrane by measuring the rate of fluorescence quenching in an instrument called a fluorescence spectrophotometer.

Although it has proven difficult to make absolute measurements of the pH and potential gradients in membrane vesicles, this type of study has been very useful for relative measurements, and for defining the energy sources and activator and inhibitor sensitivities of various transport systems. In combination with other experimental techniques, studies with membrane vesicles have provided many insights into transport in plant cells.

6

The Body System

Blood is a specialized bodily fluid that delivers necessary substances to the body's cells – such as nutrients and oxygen – and transports waste products away from those same cells.

In vertebrates, it is composed of blood cells suspended in a liquid called blood plasma. Plasma, which constitutes 55% of blood fluid, is mostly water (92% by volume), and contains dissolved proteins, glucose, mineral ions, hormones, carbon dioxide (plasma being the main medium for excretory product transportation), platelets and blood cells themselves. The blood cells present in blood are mainly red blood cells (also called RBCs or erythrocytes) and white blood cells, including leukocytes and platelets. The most abundant cells in vertebrate blood are red blood cells. These contain hemoglobin, an iron-containing protein, which facilitates transportation of oxygen by reversibly binding to this respiratory gas and greatly increasing its solubility in blood. In contrast, carbon dioxide is almost entirely transported extracellularly dissolved in plasma as bicarbonate ion.

Vertebrate blood is bright red when its hemoglobin is oxygenated. Some animals, such as crustaceans and mollusks, use hemocyanin to carry oxygen, instead of hemoglobin. Insects and some molluscs use a fluid called hemolymph instead of blood, the difference being that hemolymph is not contained in a closed circulatory system. In most insects, this "blood" does not contain oxygen-carrying molecules such as hemoglobin because their bodies are small enough for their tracheal system to suffice for supplying oxygen. Jawed vertebrates have an adaptive immune system, based largely on white blood cells. White blood cells help to resist infections and parasites. Platelets are important in the clotting of blood. Arthropods, using hemolymph, have hemocytes as part of their immune system.

Blood is circulated around the body through blood vessels by the pumping action of the heart. In animals with lungs, arterial blood carries oxygen from inhaled air to the tissues of the body, and venous blood carries carbon dioxide, a waste product of metabolism produced by cells, from the tissues to the lungs to be exhaled.

Functions

Blood performs many important functions within the body including:

- Supply of oxygen to tissues (bound to hemoglobin, which is carried in red cells)
- Supply of nutrients such as glucose, amino acids, and fatty acids (dissolved in the blood or bound to plasma proteins (e.g., blood lipids))
- Removal of waste such as carbon dioxide, urea, and lactic acid
- Immunological functions, including circulation of white blood cells, and detection of foreign material by antibodies
- Coagulation, which is one part of the body's self-repair mechanism (blood clotting after an open wound in order to stop bleeding)
- Messenger functions, including the transport of hormones and the signalling of tissue damage
- Regulation of body pH
- Regulation of core body temperature
- Hydraulic functions.

Constituents of Human Blood

Left tube: after standing, the RBCs have settled at the bottom of the tube.

Right tube: contains freshly drawn blood.

Blood accounts for 8% of the human body weight, with an average density of approximately 1060 kg/m^3, very close to pure water's density of 1000 kg/m^3. The average adult has a blood volume of roughly 5 litres (1.3 gal), composed of plasma and several kinds of cells (occasionally called *corpuscles*); these formed elements of the blood are erythrocytes (red blood cells), leukocytes (white blood cells), and thrombocytes (platelets). By volume, the red blood cells constitute about 45% of whole blood, the plasma about 54.3%, and white cells about 0.7%. Whole blood (plasma and cells) exhibits non-Newtonian fluid dynamics; its flow properties are adapted to flow effectively through tiny capillary blood vessels with less resistance than plasma by itself. In addition, if all human hemoglobin were free in the plasma rather than being contained in RBCs, the circulatory fluid would be too viscous for the cardiovascular system to function effectively.

Cells

One microliter of blood contains:

- 4.7 to 6.1 million (male), 4.2 to 5.4 million (female) erythrocytes: In most mammals, mature red blood cells lack a nucleus and organelles. They contain the blood's hemoglobin and distribute oxygen. The red blood cells (together with endothelial vessel cells and other cells) are also marked by glycoproteins that define the different blood types. The proportion of blood occupied by red blood cells is referred to as the hematocrit, and is normally about 45%. The combined surface area of all red blood cells of the human body would be roughly 2,000 times as great as the body's exterior surface.
- 4,000–11,000 leukocytes: White blood cells are part of the immune system; they destroy and remove old or aberrant cells and cellular debris, as well as attack infectious agents (pathogens) and foreign substances. The cancer of leukocytes is called leukemia.
- 200,000–500,000 thrombocytes: thrombocytes, also called platelets, are responsible for blood clotting (coagulation). They change fibrinogen into fibrin. This fibrin creates a mesh onto which red blood cells collect and clot, which then stops more blood from leaving the body and also helps to prevent bacteria from entering the body.

Plasma

About 55% of whole blood is blood plasma, a fluid that is the blood's liquid medium, which by itself is straw-yellow in colour. The blood plasma volume totals of 2.7–3.0 litres (2.8–3.2 quarts) in an average human. It is essentially an aqueous solution containing 92% water, 8% blood plasma proteins, and trace amounts of other materials. Plasma circulates dissolved nutrients, such as glucose, amino acids, and fatty acids (dissolved in the blood or bound to plasma proteins), and removes waste products, such as carbon dioxide, urea, and lactic acid.

Other important components include:

- Serum albumin
- Blood-clotting factors (to facilitate coagulation)
- Immunoglobulins (antibodies)
- Lipoprotein particles
- Various other proteins
- Various electrolytes (mainly sodium and chloride).

The term serum refers to plasma from which the clotting proteins have been removed. Most of the proteins remaining are albumin and immunoglobulins.

Narrow Range of pH Values

Blood pH is regulated to stay within the narrow range of 7.35 to 7.45, making it slightly alkaline. Blood that has a pH below 7.35 is too acidic, whereas blood pH above 7.45 is too alkaline. Blood pH, partial pressure of oxygen (pO_2), partial pressure of carbon dioxide (pCO_2), and HCO_3 are carefully regulated by a number of homeostatic mechanisms, which exert their influence principally through the respiratory system and the urinary system in order to control the acid-base balance and respiration. An arterial blood gas will measure these. Plasma also circulates hormones transmitting their messages to various tissues. The list of normal reference ranges for various blood electrolytes is extensive. Bones are especially affected by blood pH as they tend to be used as a mineral source for pH buffering. Consuming a high ratio of animal protein to vegetable protein is implicated in bone loss in women.

Blood in Non-human Vertebrates

Human blood is typical of that of mammals, although the precise details concerning cell numbers, size, protein structure, and so on, vary somewhat between species. In non-mammalian vertebrates, however, there are some key differences:

- Red blood cells of non-mammalian vertebrates are flattened and ovoid in form, and retain their cell nuclei
- There is considerable variation in the types and proportions of white blood cells; for example, acidophils are generally more common than in humans
- Platelets are unique to mammals; in other vertebrates, small, nucleated, spindle cells are responsible for blood clotting instead.

Physiology

Cardiovascular System

Blood is circulated around the body through blood vessels by the pumping action of the heart. In humans, blood is pumped from the strong left ventricle of the heart through arteries to peripheral tissues and returns to the right atrium of the heart through veins. It then enters the right ventricle and is pumped through the pulmonary artery to the lungs and returns to the left atrium through the pulmonary veins. Blood then enters the left ventricle to be circulated again. Arterial blood carries oxygen from inhaled air to all of the cells of the body, and venous blood carries carbon dioxide, a waste product of metabolism by cells, to the lungs to be exhaled. However, one exception includes pulmonary arteries, which contain the most deoxygenated blood in the body, while the pulmonary veins contain oxygenated blood.

Additional return flow may be generated by the movement of skeletal muscles, which can compress veins and push blood through the valves in veins toward the right atrium.

The blood circulation was famously described by William Harvey in 1628.

Production and Degradation of Blood Cells

In vertebrates, the various cells of blood are made in the bone marrow in a process called hematopoiesis, which includes erythropoiesis, the production of red blood cells; and myelopoiesis, the production of white blood cells and platelets. During childhood, almost every human bone produces red blood cells; as adults, red blood cell production is limited to the larger bones: the bodies of the vertebrae, the breastbone (sternum), the ribcage, the pelvic bones, and the bones of the upper arms and legs. In addition, during childhood, the thymus gland, found in the mediastinum, is an important source of lymphocytes. The proteinaceous component of blood (including clotting proteins) is produced predominantly by the liver, while hormones are produced by the endocrine glands and the watery fraction is regulated by the hypothalamus and maintained by the kidney.

Healthy erythrocytes have a plasma life of about 120 days before they are degraded by the spleen, and the Kupffer cells in the liver. The liver also clears some proteins, lipids, and amino acids. The kidney actively secretes waste products into the urine.

Oxygen Transport

About 98.5% of the oxygen in a sample of arterial blood in a healthy human breathing air at sea-level pressure is chemically combined with the Hgb. About 1.5% is physically dissolved in the other blood liquids and not connected to Hgb. The hemoglobin molecule is the primary transporter of oxygen in mammals and many other species. Hemoglobin has an oxygen binding capacity of between 1.36 and 1.37 ml O_2 per gram Hemoglobin, which increases the total blood oxygen capacity seventyfold, compared to if oxygen solely was carried by its solubility of 0.03 mL O_2 per liter blood per mmHg partial pressure of oxygen (approximately 100 mmHg in arteries).

With the exception of pulmonary and umbilical arteries and their corresponding veins, arteries carry oxygenated blood away from the heart and deliver it to the body via arterioles and capillaries, where the oxygen is consumed; afterwards, venules, and veins carry deoxygenated blood back to the heart. Under normal conditions in adult humans at rest; hemoglobin in blood leaving the lungs is about 98–99% saturated with oxygen, achieving an *oxygen delivery* of between 950 -1150 mL/min to the body. In a healthy adult at rest, *oxygen consumption* is approximately 200 -250 mL/min, and *deoxygenated* blood returning to the lungs is still approximately 75% (70 to 78%) saturated. Increased oxygen consumption during sustained exercise reduces the oxygen saturation of venous blood, which can reach less than 15% in a trained athlete; although breathing rate and blood flow increase to compensate, oxygen saturation in arterial blood can drop to 95% or less under these conditions. Oxygen saturation this low is considered dangerous in an individual at rest (for instance, during surgery under anesthesia. Sustained hypoxia (oxygenation of less than 90%), is dangerous to health, and severe hypoxia (saturations of less than 30%) may be rapidly fatal.

A fetus, receiving oxygen via the placenta, is exposed to much lower oxygen pressures (about 21% of the level found in an adult's lungs), and, so, fetuses produce another form of hemoglobin with a much higher affinity for oxygen (hemoglobin F) in order to function under these conditions.

Carbon Dioxide Transport

When blood flows through capillaries, carbon dioxide diffuses from the tissues into the blood. Some carbon dioxide is dissolved in the blood. A part of CO_2 reacts with hemoglobin and other proteins to form carbamino compounds. The remaining carbon dioxide is converted to bicarbonate and hydrogen ions through the action of RBC carbonic anhydrase. Most carbon dioxide is transported through the blood in the form of bicarbonate ions.

Carbon dioxide (CO_2), the main cellular waste product is carried in blood mainly dissolved in plasma, in equilibrium with bicarbonate (HCO_3^-) and carbonic acid (H_2CO_3). 86–90% of CO_2 in the body is converted into carbonic acid, which can quickly turn into bicarbonate, the chemical equilibrium being important in the pH buffering of plasma. Blood pH is kept in a narrow range (pH between 7.35 and 7.45).

Transport of Hydrogen Ions

Some oxyhemoglobin loses oxygen and becomes deoxyhemoglobin. Deoxyhemoglobin binds most of the hydrogen ions as it has a much greater affinity for more hydrogen than does oxyhemoglobin.

Lymphatic System

In mammals, blood is in equilibrium with lymph, which is continuously formed in tissues from blood by capillary ultrafiltration. Lymph is collected by a system of small lymphatic vessels and directed to the thoracic duct, which drains into the left subclavian vein where lymph rejoins the systemic blood circulation.

Thermoregulation

Blood circulation transports heat throughout the body, and adjustments to this flow are an important part of thermoregulation. Increasing blood flow to the surface (e.g., during warm weather or strenuous exercise) causes warmer skin, resulting in faster heat loss. In contrast, when the external temperature is low, blood flow to the extremities and surface of the skin is reduced and to prevent heat loss and is circulated to the important organs of the body, preferentially.

Hydraulic Functions

The restriction of blood flow can also be used in specialized tissues to cause engorgement, resulting in an erection of that tissue; examples are the erectile tissue in the penis and clitoris. Another example of a hydraulic function is the jumping spider, in which blood forced into the legs under pressure causes them to straighten for a powerful jump, without the need for bulky muscular legs.

Invertebrates

In insects, the blood (more properly called hemolymph) is not involved in the transport of oxygen. (Openings called tracheae allow oxygen from the air to diffuse directly to the tissues). Insect blood moves nutrients to the tissues and removes waste products in an open system.

Other invertebrates use respiratory proteins to increase the oxygen-carrying capacity. Hemoglobin is the most common respiratory protein found in nature. Hemocyanin (blue) contains copper and is found in crustaceans and mollusks. It is thought that tunicates (sea squirts) might use vanabins (proteins containing vanadium) for respiratory pigment (bright-green, blue, or orange). In many invertebrates, these oxygen-carrying proteins are freely soluble in the blood; in vertebrates they are contained in specialized red blood cells, allowing for a higher concentration of respiratory pigments without increasing viscosity or damaging blood filtering organs like the kidneys. Giant tube worms have unusual hemoglobins that allow them to live in extraordinary environments. These hemoglobins also carry sulfides normally fatal in other animals.

Colour

Hemoglobin

Hemoglobin is the principal determinant of the colour of blood in vertebrates. Each molecule has four heme groups, and their interaction with various molecules alters the exact colour. In vertebrates and other hemoglobin-using creatures, arterial blood and capillary blood are bright red, as oxygen imparts a strong red colour to the heme group. Deoxygenated blood is a darker shade of red with a bluish tinge; this is present in veins, and can be seen during blood donation and when venous blood samples are taken.

Blood in carbon monoxide poisoning is bright red, because carbon monoxide causes the formation of carboxyhemoglobin. In cyanide poisoning, the body cannot utilize oxygen, so the venous blood remains oxygenated, increasing the redness. While hemoglobin-containing blood is never blue, there are several conditions and diseases wherein the colour of the heme groups make the skin appear blue. If the heme is oxidized, methaemoglobin, which is more brownish and cannot transport oxygen, is formed. In the rare condition sulfhemoglobinemia, arterial hemoglobin is partially oxygenated, and appears dark red with a bluish hue (cyanosis).

Veins in the skin appear blue for a variety of reasons only weakly dependent on the colour of the blood. Light scattering in the skin, and the visual processing of colour play roles as well. Skinks in the genus *Prasinohaema* have green blood due to a buildup of the waste product biliverdin.

Hemocyanin

The blood of most mollusks – including cephalopods and gastropods – as well as some arthropods, such as horseshoe crabs, is blue, as it contains the copper-containing protein hemocyanin at concentrations of about 50 grams per liter. Hemocyanin is colorless when deoxygenated and dark blue when oxygenated. The blood in the circulation of these creatures, which generally live in cold environments with low oxygen tensions, is grey-white to pale yellow, and it turns dark blue when exposed to the oxygen in the air, as seen when they bleed. This is due to change in colour of hemocyanin when it is oxidized. Hemocyanin carries oxygen in extracellular fluid, which is in contrast to the intracellular oxygen transport in mammals by hemoglobin in RBCs.

Hemovanabin

The blood of some species of ascidians and tunicates, also known as sea squirts and sea cucumbers, contains proteins called Vanabins. These proteins are based on vanadium, and give the creatures a concentration of vanadium in their bodies 100 times higher than the surrounding sea water. It is not clear whether these vanabins actually carry oxygen. When exposed to oxygen, however, vanabins turn a mustard yellow.

Pathology

General Medical Disorders

- Disorders of volume
  - Injury can cause blood loss through bleeding. A healthy adult can lose almost 20% of blood volume (1 L) before the first symptom, restlessness, begins, and 40% of volume (2 L) before shock sets in. Thrombocytes are important for blood coagulation

and the formation of blood clots, which can stop bleeding. Trauma to the internal organs or bones can cause internal bleeding, which can sometimes be severe.

- o Dehydration can reduce the blood volume by reducing the water content of the blood. This would rarely result in shock (apart from the very severe cases) but may result in orthostatic hypotension and fainting.

- Disorders of circulation
  - o Shock is the ineffective perfusion of tissues, and can be caused by a variety of conditions including blood loss, infection, poor cardiac output.
  - o Atherosclerosis reduces the flow of blood through arteries, because atheroma lines arteries and narrows them. Atheroma tends to increase with age, and its progression can be compounded by many causes including smoking, high blood pressure, excess circulating lipids (hyperlipidemia), and diabetes mellitus.
  - o Coagulation can form a thrombosis, which can obstruct vessels.
  - o Problems with blood composition, the pumping action of the heart, or narrowing of blood vessels can have many consequences including hypoxia (lack of oxygen) of the tissues supplied. The term *ischemia* refers to tissue that is inadequately perfused with blood, and *infarction* refers to tissue death (necrosis), which can occur when the blood supply has been blocked (or is very inadequate).

Hematological Disorders

- Anemia
  - o Insufficient red cell mass (anemia) can be the result of bleeding, blood disorders like thalassemia, or nutritional deficiencies; and may require blood transfusion. Several countries have blood banks to fill the demand for transfusable blood. A person receiving a blood transfusion must have a blood type compatible with that of the donor.
  - o Sickle-cell anemia
- Disorders of cell proliferation
  - o Leukemia is a group of cancers of the blood-forming tissues.
  - o Non-cancerous overproduction of red cells (polycythemia vera) or platelets (essential thrombocytosis) may be premalignant.
  - o Myelodysplastic syndromes involve ineffective production of one or more cell lines.

- Disorders of coagulation
  - o Hemophilia is a genetic illness that causes dysfunction in one of the blood's clotting mechanisms. This can allow otherwise inconsequential wounds to be life-threatening, but more commonly results in hemarthrosis, or bleeding into joint spaces, which can be crippling.
  - o Ineffective or insufficient platelets can also result in coagulopathy (bleeding disorders).
  - o Hypercoagulable state (thrombophilia) results from defects in regulation of platelet or clotting factor function, and can cause thrombosis.
- Infectious disorders of blood
  - o Blood is an important vector of infection. HIV, the virus, which causes AIDS, is transmitted through contact with blood, semen or other body secretions of an infected person. Hepatitis B and C are transmitted primarily through blood contact. Owing to blood-borne infections, bloodstained objects are treated as a biohazard.
  - o Bacterial infection of the blood is bacteremia or sepsis. Viral Infection is viremia. Malaria and trypanosomiasis are blood-borne parasitic infections.

Carbon Monoxide Poisoning

Substances other than oxygen can bind to hemoglobin; in some cases this can cause irreversible damage to the body. Carbon monoxide, for example, is extremely dangerous when carried to the blood via the lungs by inhalation, because carbon monoxide irreversibly binds to hemoglobin to form carboxyhemoglobin, so that less hemoglobin is free to bind oxygen, and fewer oxygen molecules can be transported throughout the blood. This can cause suffocation insidiously. A fire burning in an enclosed room with poor ventilation presents a very dangerous hazard, since it can create a build-up of carbon monoxide in the air. Some carbon monoxide binds to hemoglobin when smoking tobacco.

Carbon monoxide poisoning occurs after enough inhalation of carbon monoxide (CO). Carbon monoxide is a toxic gas, but, being colorless, odorless, tasteless, and non-irritating, it is very difficult for people to detect. Carbon monoxide is a product of incomplete combustion of organic matter with insufficient oxygen supply to enable complete oxidation to carbon dioxide (CO_2) and is often produced in domestic or industrial settings by motor vehicles and other gasoline-powered tools, heaters, and cooking equipment. Exposures at 100 ppm or greater can be dangerous to human health.

Symptoms of mild acute poisoning include headaches, vertigo, and flu-like effects; larger exposures can lead to significant toxicity of the central nervous system and heart, and even death. Following acute poisoning, long-term sequelae often occur. Carbon monoxide can also have severe effects on the fetus of a pregnant woman. Chronic exposure to low levels of carbon monoxide can lead to depression, confusion, and memory loss. Carbon monoxide mainly causes adverse effects in humans by combining with hemoglobin to form carboxyhemoglobin (HbCO) in the blood. This prevents oxygen binding to hemoglobin, reducing the oxygen-carrying capacity of the blood, leading to hypoxia. Additionally, myoglobin and mitochondrial cytochrome oxidase are thought to be adversely affected. Carboxyhemoglobin can revert to hemoglobin, but the recovery takes time because the HbCO complex is fairly stable.

Acute Poisoning

The main manifestations of poisoning develop in the organ systems most dependent on oxygen use, the central nervous system and the heart. The initial symptoms of acute carbon monoxide poisoning include headache, nausea, malaise, and fatigue. These symptoms are often mistaken for a virus such as influenza or other illnesses such as food poisoning or gastroenteritis.

Headache is the most common symptom of acute carbon monoxide poisoning; it is often described as dull, frontal, and continuous. Increasing exposure produces cardiac abnormalities include fast heart rate, low blood pressure, and cardiac arrhythmia; central nervous system symptoms include delirium, hallucinations, dizziness, unsteady gait, confusion, seizures, central nervous system depression, unconsciousness, respiratory arrest, and even death. Less common symptoms of acute carbon monoxide poisoning include myocardial ischemia, atrial fibrillation, pneumonia, pulmonary edema, high blood sugar, lactic acidosis, muscle necrosis, acute kidney failure, skin lesions, and visual and auditory problems.

One of the major concerns following acute carbon monoxide poisoning is the severe delayed neurological manifestations that may occur. Problems may include difficulty with higher intellectual functions, short-term memory loss, dementia, amnesia, psychosis, irritability, a strange gait, speech disturbances, parkinson's disease-like syndromes, cortical blindness, and a depressed mood. Depression may even occur in those who did not have pre-existing depression. These delayed neurological sequelae may occur in up to 50% of poisoned patients after 2 to 40 days. It is difficult to predict who will develop delayed sequelae; however, advancing age, loss of consciousness while poisoned, and initial neurological abnormalities may increase the chance of developing delayed symptoms.

Chronic Poisoning

Chronic exposure to relatively low levels of carbon monoxide may cause persistent headaches, lightheadedness, depression, confusion, memory loss, nausea and vomiting. It is unknown whether low-level chronic exposure may cause permanent neurological damage.

Typically, upon removal from exposure to carbon monoxide, symptoms usually resolve themselves, unless there has been an episode of severe acute poisoning. However, one case noted permanent memory loss and learning problems after a 3-year exposure to relatively low levels of carbon monoxide from a faulty furnace. Chronic exposure may worsen cardiovascular symptoms in some people. Long-term exposures to carbon monoxide present the greatest risk to persons with coronary heart disease and in people who are pregnant.

Causes

Concentration	*Source*
0.1 ppm	Natural atmosphere level (MOPITT)
0.5 to 5 ppm	Average level in homes
5 to 15 ppm	Near properly adjusted gas stoves in homes
100 to 200 ppm	Exhaust from automobiles in the Mexico City central area
5,000 ppm	Exhaust from a home wood fire
7,000 ppm	Undiluted warm car exhaust without a catalytic converter

Carbon monoxide is a product of combustion of organic matter under conditions of restricted oxygen supply, which prevents complete oxidation to carbon dioxide (CO_2). Sources of carbon monoxide include house fires, faulty furnaces, heaters, wood-burning stoves, internal combustion vehicle exhaust, electrical generators, propane-fuelled equipment such as portable stoves, and gasoline-powered tools such as leaf blowers, lawn mowers, high-pressure washers, concrete cutting saws, power trowels, and welders. Exposure typically occurs when equipment is used in buildings or semi-enclosed spaces. Poisoning may also occur following the use of a self-contained underwater breathing apparatus (SCUBA) due to faulty diving air compressors.

Riding in pickup trucks has led to poisoning in children. Idling automobiles with the exhaust pipe blocked by snow has led to the poisoning of car occupants. Generators and propulsion engines on boats, especially houseboats, has resulted in fatal carbon monoxide exposures. Another source of poisoning is exposure to the organic solvent dichloromethane, found in some paint strippers. Dichloromethane is converted into carbon monoxide by the body. In some caves carbon monoxide can build up in enclosed chambers due to decomposing organic matter.

Pathophysiology

The precise mechanisms by which the effects of carbon monoxide are induced upon bodily systems, are complex and not yet fully understood. Known mechanisms include carbon monoxide binding to hemoglobin, myoglobin and mitochondrial cytochrome oxidase, and carbon monoxide causing brain lipid peroxidation.

Hemoglobin

Following absorption, carbon monoxide binds to hemoglobin, which is the principal oxygen-carrying compound in blood; this produces a compound known as carboxyhemoglobin. The traditional belief is that carbon monoxide toxicity arises from the formation of carboxyhemoglobin, which decreases the oxygen-carrying capacity of the blood and inhibits the transport, delivery, and utilization of oxygen by the body. The affinity between hemoglobin and carbon monoxide is approximately 230 times stronger than the affinity between hemoglobin and oxygen so carbon monoxide binds to hemoglobin in preference to oxygen. Hemoglobin is a tetramer with four oxygen binding sites. The binding of carbon monoxide at one of these sites increases the oxygen affinity of the remaining three sites, which causes the hemoglobin molecule to retain oxygen that would otherwise be delivered to the tissue. This situation is described as carbon monoxide shifting the oxygen dissociation curve to the left. Because of the increased affinity between hemoglobin and oxygen during carbon monoxide poisoning, the blood oxygen content is increased. But because all the oxygen stays in the hemoglobin, none is delivered to the tissues. This causes hypoxic tissue injury. Hemoglobin acquires a bright red colour when converted into carboxyhemoglobin, so poisoned patients have been described as looking pink-cheeked and healthy. However, this cherry-red appearance is rarely seen in living patients so is not considered a reliable diagnostic sign.

Myoglobin

Carbon monoxide also binds to the hemeprotein myoglobin. It has a high affinity for myoglobin, about 60 times greater than that of oxygen. Carbon monoxide bound to myoglobin may impair its ability to utilize oxygen. This causes reduced cardiac output and hypotension, which may result in brain ischemia. A delayed return of symptoms have been reported. This results following a recurrence of increased carboxyhemoglobin levels; this effect may be due to a late release of carbon monoxide from myoglobin, which subsequently binds to hemoglobin.

Cytochrome Oxidase

Another mechanism involves effects on the mitochondrial respiratory enzyme chain that is responsible for effective tissue utilization of oxygen.

Carbon monoxide binds to cytochrome oxidase with less affinity than oxygen, so it is possible that it requires significant intracellular hypoxia before binding. This binding interferes with aerobic metabolism and efficient adenosine triphosphate synthesis. Cells respond by switching to anaerobic metabolism, causing anoxia, lactic acidosis, and eventual cell death. The rate of dissociation between carbon monoxide and cytochrome oxidase is slow, causing a relatively prolonged impairment of oxidative metabolism.

Central Nervous System Effects

The mechanism that is thought to have a significant influence on delayed effects involves formed blood cells and chemical mediators, which cause brain lipid peroxidation (degradation of unsaturated fatty acids). Carbon monoxide causes endothelial cell and platelet release of nitric oxide, and the formation of oxygen free radicals including peroxynitrite. In the brain this causes further mitochondrial dysfunction, capillary leakage, leukocyte sequestration, and apoptosis. The result of these effects is lipid peroxidation, which causes delayed reversible demyelinization of white matter in the central nervous system, which can lead to edema and necrosis within the brain. This brain damage occurs mainly during the recovery period. This may result in cognitive defects, especially affecting memory and learning, and movement disorders. These disorders are typically related to damage to the cerebral white matter and basal ganglia. Hallmark pathological changes following poisoning is bilateral necrosis of the white matter, globus pallidus, cerebellum, hippocampus and the cerebral cortex.

Pregnancy

Carbon monoxide poisoning in pregnant women may cause severe adverse fetal effects. Poisoning causes fetal tissue hypoxia by decreasing the release of maternal oxygen to the fetus. Carbon monoxide also crosses the placenta and combines with fetal hemoglobin, causing more direct fetal tissue hypoxia. Additionally, fetal hemoglobin has a 10 to 15% higher affinity for carbon monoxide than adult hemoglobin, causing more severe poisoning in the fetus than in the adult. Elimination of carbon monoxide is slower in the fetus, leading to an accumulation of the toxic chemical. The level of fetal morbidity and mortality in acute carbon monoxide poisoning is significant, so despite mild maternal poisoning or following maternal recovery, severe fetal poisoning or death may still occur.

Excretion

Excretion is the process by which waste products of metabolism and other non-useful materials are eliminated from an organism. It is an essential process in all forms of life. It contrasts secretion, where the substance may have specific tasks after leaving the cell.

In single-celled organisms, waste products are discharged directly through the surface of the cell. Multicellular organisms utilize more complex excretory methods. Higher plants eliminate gases through the stomata, or pores, on the surface of leaves. Animals have special excretory organs.

Non-mammalian

In plants, breakdown of substances is much slower than in animals. Hence accumulation of waste is much slower and there are no special organs of excretion. Green plants in darkness or plants that do not contain chlorophyll produce carbon dioxide and water as respiratory waste products. Carbon dioxide released during respiration gets utilized during photosynthesis. Oxygen itself can be thought of as a waste product generated during photosynthesis, and exits through stomata, root cell walls, and other routes. Plants can get rid of excess water by transpiration and guttation. Waste products may be stored in leaves that fall off. Other waste materials that are exuded by some plants — resins, saps, latexes, etc. are forced from the interior of the plant by hydrostatic pressures inside the plant and by absorptive forces of plant cells. These processes do not need added energy, they act passively. Plants also excrete some waste substances into the soil around them.

Aquatic animals usually excrete ammonia directly into the external environment, as this compound has high solubility and there is ample water available for dilution. In terrestrial animals ammonia-like compounds are converted into other nitrogenous materials as there is less water in the environment and ammonia itself is toxic. Birds excrete their nitrogenous wastes as uric acid in the form of a paste. This is metabolically more expensive, but allows more efficient water retention and it can be stored more easily in the egg. Many avian species, especially seabirds, can also excrete salt via specialized nasal salt glands, the saline solution leaving through nostrils in the beak. In insects, a system involving Malpighian tubules is utilized to excrete metabolic waste. Metabolic waste diffuses or is actively transported into the tubule, which transports the wastes to the intestines. The metabolic waste is then released from the body along with fecal matter.

Many people misuse the term excretion as a euphemism for defecation, and use excrement for feces, but this is biologically incorrect.

Nervous System

The nervous system is an organ system containing a network of specialized cells called neurons that coordinate the actions of an animal and transmit signals between different parts of its body. In most animals the nervous system consists of two parts, central and peripheral. The central nervous system of vertebrates (such as humans) contains the brain, spinal cord, and retina. The peripheral nervous system consists of sensory neurons, clusters of neurons called ganglia, and nerves connecting them to each other and to the central nervous system. These regions are all interconnected by means of complex neural pathways. The enteric nervous system, a subsystem of the peripheral nervous system, has the capacity, even when severed from the rest of the nervous system through its primary connection by the vagus nerve, to function independently in controlling the gastrointestinal system.

Neurons send signals to other cells as electrochemical waves travelling along thin fibers called axons, which cause chemicals called neurotransmitters to be released at junctions called synapses. A cell that receives a synaptic signal may be excited, inhibited, or otherwise modulated. Sensory neurons are activated by physical stimuli impinging on them, and send signals that inform the central nervous system of the state of the body and the external environment. Motor neurons, situated either in the central nervous system or in peripheral ganglia, connect the nervous system to muscles or other effector organs. Central neurons, which in vertebrates greatly outnumber the other types, make all of their input and output connections with other neurons. The interactions of all these types of neurons form neural circuits that generate an organism's perception of the world and determine its behaviour. Along with neurons, the nervous system contains other specialized cells called glial cells (or simply glia), which provide structural and metabolic support.

Nervous systems are found in most multicellular animals, but vary greatly in complexity. Sponges have no nervous system, although they have homologs of many genes that play crucial roles in nervous system function, and are capable of several whole-body responses, including a primitive form of locomotion. Placozoans and mesozoans—other simple animals that are not classified as part of the subkingdom Eumetazoa—also have no nervous system. In Radiata (radially symmetric animals such as jellyfish) the nervous system consists of a simple nerve net. Bilateria, which include the great majority of vertebrates and invertebrates, all have a nervous system containing a brain, one central cord (or two running in parallel), and peripheral nerves. The size of the bilaterian nervous system ranges from a few hundred cells in the simplest worms, to on the order of 100 billion cells in humans.

Structure

The nervous system derives its name from nerves, which are cylindrical bundles of tissue that emanate from the brain and central cord, and branch repeatedly to innervate every part of the body. Nerves are large enough to have been recognized by the ancient Egyptians, Greeks, and Romans, but their internal structure was not understood until it became possible to examine them using a microscope. A microscopic examination shows that nerves consist primarily of the axons of neurons, along with a variety of membranes that wrap around them and segregate them into fascicles. The neurons that give rise to nerves do not lie within them—their cell bodies reside within the brain, central cord, or peripheral ganglia.

All animals more advanced than sponges have nervous systems. However, even sponges, unicellular animals, and non-animals such as slime molds have cell-to-cell signalling mechanisms that are precursors to those of neurons. In radially symmetric animals such as the jellyfish and hydra, the nervous system consists of a diffuse network of isolated cells. In bilaterian animals, which make up the great majority of existing species, the nervous system has a common structure that originated early in the Cambrian period, over 500 million years ago.

Cells

The nervous system is primarily made up of two categories of cells: neurons and glial cells.

Neurons

The nervous system is defined by the presence of a special type of cell—the neuron (sometimes called "neurone" or "nerve cell"). Neurons can be distinguished from other cells in a number of ways, but their most fundamental property is that they communicate with other cells via synapses, which are membrane-to-membrane junctions containing molecular machinery that allows rapid transmission of signals, either electrical or chemical. Many types of neuron possess an axon, a protoplasmic protrusion that can extend to distant parts of the body and make thousands of synaptic contacts. Axons frequently travel through the body in bundles called nerves.

Even in the nervous system of a single species such as humans, hundreds of different types of neurons exist, with a wide variety of morphologies and functions. These include sensory neurons that transmute physical stimuli such as light and sound into neural signals, and motor neurons that transmute neural signals into activation of muscles or glands; however in many species the great majority of neurons receive all of their input from other neurons and send their output to other neurons.

Glial Cells

Glial cells are non-neuronal cells that provide support and nutrition, maintain homeostasis, form myelin, and participate in signal transmission in the nervous system. In the human brain, it is estimated that the total number of glia roughly equals the number of neurons, although the proportions vary in different brain areas. Among the most important functions of glial cells are to support neurons and hold them in place; to supply nutrients to neurons; to insulate neurons electrically; to destroy pathogens and remove dead neurons; and to provide guidance cues directing the axons of neurons to their targets. A very important type of glial cell (oligodendrocytes in the central nervous system, and Schwann cells in the peripheral nervous system) generates layers of a fatty substance called myelin that wraps around axons and provides electrical insulation which allows them to transmit action potentials much more rapidly and efficiently.

Anatomy in Vertebrates

The nervous system of vertebrate animals (including humans) is divided into the central nervous system (CNS) and peripheral nervous system (PNS).

The central nervous system (CNS) is the largest part, and includes the brain and spinal cord. The spinal cavity contains the spinal cord, while the head contains the brain. The CNS is enclosed and protected by meninges, a three-layered system of membranes, including a tough, leathery outer layer called the dura mater. The brain is also protected by the skull, and the spinal cord by the vertebrae.

The peripheral nervous system (PNS) is a collective term for the nervous system structures that do not lie within the CNS. The large majority of the axon bundles called nerves are considered to belong to the PNS, even when the cell bodies of the neurons to which they belong reside within the brain or spinal cord. The PNS is divided into somatic and visceral parts. The somatic part consists of the nerves that innervate the skin, joints, and muscles. The cell bodies of somatic sensory neurons lie in dorsal root ganglia of the spinal cord. The visceral part, also known as the autonomic nervous system, contains neurons that innervate the internal organs, blood vessels, and glands. The autonomic nervous system itself consists of two parts: the sympathetic nervous system and the parasympathetic nervous system. Some authors also include sensory neurons whose cell bodies lie in the periphery (for senses such as hearing) as part of the PNS; others, however, omit them.

The vertebrate nervous system can also be divided into areas called grey matter ("gray matter" in American spelling) and white matter. Grey matter (which is only grey in preserved tissue, and is better described as pink or light brown in living tissue) contains a high proportion of cell bodies of neurons. White matter is composed mainly of myelinated axons, and takes its colour from the myelin. White matter includes all of the peripheral nerves, and much of the interior of the brain and spinal cord.

Grey matter is found in clusters of neurons in the brain and spinal cord, and in cortical layers that line their surfaces. There is an anatomical convention that a cluster of neurons in the brain or spinal cord is called a nucleus, whereas a cluster of neurons in the periphery is called a ganglion. There are, however, a few exceptions to this rule, notably including the part of the forebrain called the basal ganglia.

Comparative Anatomy and Evolution

Neural Precursors in Sponges

Sponges have no cells connected to each other by synaptic junctions, that is, no neurons, and therefore no nervous system. They do, however, have homologs of many genes that play key roles in synaptic function. Recent studies have shown that sponge cells express a group of proteins that cluster together to form a structure resembling a postsynaptic density (the signal-receiving part of a synapse). However, the function of this structure is currently unclear. Although sponge cells do not show synaptic transmission, they do communicate with each other via calcium waves and other impulses, which mediate some simple actions such as whole-body contraction.

Radiata

Jellyfish, comb jellies, and related animals have diffuse nerve nets rather than a central nervous system. In most jellyfish the nerve net is spread more or less evenly across the body; in comb jellies it is concentrated near the mouth. The nerve nets consist of sensory neurons that pick up chemical, tactile, and visual signals, motor neurons that can activate contractions of the body wall, and intermediate neurons that detect patterns of activity in the sensory neurons and send signals to groups of motor neurons as a result. In some cases groups of intermediate neurons are clustered into discrete ganglia.

The development of the nervous system in radiata is relatively unstructured. Unlike bilaterians, radiata only have two primordial cell layers, endoderm and ectoderm. Neurons are generated from a special set of ectodermal precursor cells, which also serve as precursors for every other ectodermal cell type.

Bilateria

The vast majority of existing animals are bilaterians, meaning animals with left and right sides that are approximate mirror images of each other. All bilateria are thought to have descended from a common wormlike ancestor that appeared in the Cambrian period, 550–600 million years ago. The fundamental bilaterian body form is a tube with a hollow gut cavity running from mouth to anus, and a nerve cord with an enlargement (a "ganglion") for each body segment, with an especially large ganglion at the front, called the "brain".

Even mammals, including humans, show the segmented bilaterian body plan at the level of the nervous system. The spinal cord contains a series of segmental ganglia, each giving rise to motor and sensory nerves that innervate a portion of the body surface and underlying musculature. On the limbs, the layout of the innervation pattern is complex, but on the trunk it gives rise to a series of narrow bands. The top three segments belong to the brain, giving rise to the forebrain, midbrain, and hindbrain.

Bilaterians can be divided, based on events that occur very early in embryonic development, into two groups (superphyla) called protostomes and deuterostomes. Deuterostomes include vertebrates as well as echinoderms, hemichordates (mainly acorn worms), and Xenoturbellidans. Protostomes, the more diverse group, include arthropods, molluscs, and numerous types of worms. There is a basic difference between the two groups in the placement of the nervous system within the body: protostomes possess a nerve cord on the ventral (usually bottom) side of the body, whereas in deuterostomes the nerve cord is on the dorsal (usually top) side. In fact, numerous aspects of the body are inverted between the two groups, including the expression patterns of several genes that show dorsal-to-ventral gradients. Most anatomists now consider that the bodies of protostomes and deuterostomes are "flipped over" with respect to each other, a hypothesis that was first proposed by Geoffroy Saint-Hilaire for insects in comparison to vertebrates. Thus insects, for example, have nerve cords that run along the ventral midline of the body, while all vertebrates have spinal cords that run along the dorsal midline.

Worms

Worms are the simplest bilaterian animals, and reveal the basic structure of the bilaterian nervous system in the most straightforward way. As an example, earthworms have dual nerve cords running along the length of the body and merging at the tail and the mouth. These nerve cords are connected by transverse nerves like the rungs of a ladder. These transverse nerves help coordinate the two sides of the animal. Two ganglia at the head end function similar to a simple brain. Photoreceptors on the animal's eyespots provide sensory information on light and dark.

The nervous system of one very small worm, the roundworm *Caenorhabditis elegans*, has been mapped out down to the synaptic level. Every neuron and its cellular lineage has been recorded and most, if not all, of the neural connections are known. In this species, the nervous system is sexually dimorphic; the nervous systems of the two sexes, males and hermaphrodites, have different numbers of neurons and groups of neurons that perform sex-specific functions. In *C. elegans*, males have exactly 383 neurons, while hermaphrodites have exactly 302 neurons.

Arthropods

Arthropods, such as insects and crustaceans, have a nervous system made up of a series of ganglia, connected by a ventral nerve cord made up of two parallel connectives running along the length of the belly. Typically, each body segment has one ganglion on each side, though some ganglia are fused to form the brain and other large ganglia. The head segment contains the brain, also known as the supraesophageal ganglion. In the insect nervous system, the brain is anatomically divided into the protocerebrum, deutocerebrum, and tritocerebrum. Immediately behind the brain is the subesophageal ganglion, which is composed of three pairs of fused ganglia. It controls the mouthparts, the salivary glands and certain muscles. Many arthropods have well-developed sensory organs, including compound eyes for vision and antennae for olfaction and pheromone sensation. The sensory information from these organs is processed by the brain. In insects, many neurons have cell bodies that are positioned at the edge of the brain and are electrically passive—the cell bodies serve only to provide metabolic support and do not participate in signalling. A protoplasmic fiber runs from the cell body and branches profusely, with some parts transmitting signals and other parts receiving signals. Thus, most parts of the insect brain have passive cell bodies arranged around the periphery, while the neural signal processing takes place in a tangle of protoplasmic fibers called neuropil, in the interior.

"Identified" Neurons

A neuron is called *identified* if it has properties that distinguish it from every other neuron in the same animal—properties such as location, neurotransmitter, gene expression pattern, and connectivity—and if every individual organism belonging to the same species has one and only one neuron with the same set of properties. In vertebrate nervous systems very few neurons are "identified" in this sense—in humans, there are believed to be none—but in simpler nervous systems, some or all neurons may be thus unique. In the roundworm C. elegans, whose nervous system is the most thoroughly described of any animal's, every neuron in the body is uniquely identifiable, with the same location and the same connections in every individual worm. One notable consequence of this fact is that the form of the *C. elegans* nervous system is completely specified by the genome, with no experience-dependent plasticity.

The brains of many molluscs and insects also contain substantial numbers of identified neurons. In vertebrates, the best known identified neurons are the gigantic Mauthner cells of fish. Every fish has two Mauthner cells, located in the bottom part of the brainstem, one on the left side and one on the right. Each Mauthner cell has an axon that crosses over, innervating neurons at the same brain level and then travelling down through the spinal cord, making numerous connections as it goes. The synapses generated by a Mauthner cell are so powerful that a single action potential gives rise to a major behavioral response: within milliseconds the fish curves its body into a C-shape, then straightens, thereby propelling itself rapidly forward. Functionally this is a fast escape response, triggered most easily by a strong sound wave or pressure wave impinging on the lateral line organ of the fish. Mauther cells are not the only identified neurons in fish—there are about 20 more types, including pairs of "Mauthner cell analogs" in each spinal segmental nucleus. Although a Mauthner cell is capable of bringing about an escape response all by itself, in the context of ordinary behaviour other types of cells usually contribute to shaping the amplitude and direction of the response.

Mauthner cells have been described as command neurons. A command neuron is a special type of identified neuron, defined as a neuron that is capable of driving a specific behaviour all by itself. Such neurons appear most commonly in the fast escape systems of various species—the squid giant axon and squid giant synapse, used for pioneering experiments in neurophysiology because of their enormous size, both participate in the fast escape circuit of the squid. The concept of a command neuron has, however, become controversial, because of studies showing that some neurons that initially appeared to fit the description were really only capable of evoking a response in a limited set of circumstances.

Function

At the most basic level, the function of the nervous system is to send signals from one cell to others, or from one part of the body to others. There are multiple ways that a cell can send signals to other cells. One is by releasing chemicals called hormones into the internal circulation, so that they can diffuse to distant sites. In contrast to this "broadcast" mode of signalling, the nervous system provides "point-to-point" signals—neurons project their axons to specific target areas and make synaptic connections with specific target cells. Thus, neural signalling is capable of a much higher level of specificity than hormonal signalling. It is also much faster: the fastest nerve signals travel at speeds that exceed 100 meters per second.

At a more integrative level, the primary function of the nervous system is to control the body. It does this by extracting information from the environment using sensory receptors, sending signals that encode this information into the central nervous system, processing the information to determine an appropriate response, and sending output signals to muscles or glands to activate the response. The evolution of a complex nervous system has made it possible for various animal species to have advanced perception abilities such as vision, complex social interactions, rapid coordination of organ systems, and integrated processing of concurrent signals. In humans, the sophistication of the nervous system makes it possible to have language, abstract representation of concepts, transmission of culture, and many other features of human society that would not exist without the human brain.

Neurons and Synapses

Most neurons send signals via their axons, although some types are capable of dendrite-to-dendrite communication. (In fact, the types of neurons called amacrine cells have no axons, and communicate only via their dendrites.) Neural signals propagate along an axon in the form of electrochemical waves called action potentials, which produce cell-to-cell signals at points where axon terminals make synaptic contact with other cells.

Synapses may be electrical or chemical. Electrical synapses make direct electrical connections between neurons, but chemical synapses are much more common, and much more diverse in function. At a chemical synapse, the cell that sends signals is called presynaptic, and the cell that receives signals is called postsynaptic. Both the presynaptic and postsynaptic areas are full of molecular machinery that carries out the signalling process. The presynaptic area contains large numbers of tiny spherical vessels called synaptic vesicles, packed with neurotransmitter chemicals. When the presynaptic terminal is electrically stimulated, an array of molecules embedded in the membrane are activated, and cause the contents of the vesicles to be released into the narrow space between the presynaptic and postsynaptic membranes, called the synaptic cleft. The neurotransmitter then binds to receptors embedded in the postsynaptic membrane, causing them to enter an activated state. Depending on the type of receptor, the resulting effect on the postsynaptic cell may be excitatory, inhibitory, or modulatory in more complex ways. For example, release of the neurotransmitter acetylcholine at a synaptic contact between a motor neuron and a muscle cell induces rapid contraction of the muscle cell. The entire synaptic transmission process takes only a fraction of a millisecond, although the effects on the postsynaptic cell may last much longer (even indefinitely, in cases where the synaptic signal leads to the formation of a memory trace).

There are literally hundreds of different types of synapses. In fact, there are over a hundred known neurotransmitters, and many of them have multiple types of receptor. Many synapses use more than one neurotransmitter—a common arrangement is for a synapse to use one fast-acting small-molecule neurotransmitter such as glutamate or GABA, along with one or more peptide neurotransmitters that play slower-acting modulatory roles. Molecular neuroscientists generally divide receptors into two broad groups: chemically gated ion channels and second messenger systems. When a chemically gated ion channel is activated, it forms a passage that allow specific types of ion to flow across the membrane. Depending on the type of ion, the effect on the target cell may be excitatory or inhibitory. When a second messenger system is activated, it starts a cascade of molecular interactions inside the target cell, which may ultimately produce a wide variety of complex effects, such as increasing or decreasing the sensitivity of the cell to stimuli, or even altering gene transcription.

According to a rule called Dale's principle, which has only a few known exceptions, a neuron releases the same neurotransmitters at all of its synapses. This does not mean, though, that a neuron exerts the same effect on all of its targets, because the effect of a synapse depends not on the neurotransmitter, but on the receptors that it activates. Because different targets can (and frequently do) use different types of receptors, it is possible for a neuron to have excitatory effects on one set of target cells, inhibitory effects on others, and complex modulatory effects on others still. Nevertheless, it happens that the two most widely used neurotransmitters, glutamate and GABA, each have largely consistent effects. Glutamate has several widely occurring types of receptors, but all of them are excitatory or modulatory. Similarly, GABA has several widely occurring receptor types, but all of them are inhibitory. Because of this consistency, glutamatergic cells are frequently referred to as "excitatory neurons", and GABAergic cells as "inhibitory neurons". Strictly speaking this is an abuse of terminology—it is the receptors that are excitatory and inhibitory, not the neurons—but it is commonly seen even in scholarly publications.

One very important subset of synapses are capable of forming memory traces by means of long-lasting activity-dependent changes in synaptic strength. The best-known form of neural memory is a process called long-term potentiation (abbreviated LTP), which operates at synapses that use the neurotransmitter glutamate acting on a special type of receptor known as the NMDA receptor.

The NMDA receptor has an "associative" property: if the two cells involved in the synapse are both activated at approximately the same time, a channel opens that permits calcium to flow into the target cell. The calcium entry initiates a second messenger cascade that ultimately leads to an increase in the number of glutamate receptors in the target cell, thereby increasing the effective strength of the synapse. This change in strength can last for weeks or longer. Since the discovery of LTP in 1973, many other types of synaptic memory traces have been found, involving increases or decreases in synaptic strength that are induced by varying conditions, and last for variable periods of time. Reward learning, for example, depends on a variant form of LTP that is conditioned on an extra input coming from a reward-signalling pathway that uses dopamine as neurotransmitter. All these forms of synaptic modifiability, taken collectively, give rise to neural plasticity, that is, to a capability for the nervous system to adapt itself to variations in the environment.

Neural Circuits and Systems

The basic neuronal function of sending signals to other cells includes a capability for neurons to exchange signals with each other. Networks formed by interconnected groups of neurons are capable of a wide variety of functions, including feature detection, pattern generation, and timing. In fact, it is difficult to assign limits to the types of information processing that can be carried out by neural networks: Warren McCulloch and Walter Pitts showed in 1943 that even networks formed from a greatly simplified mathematical abstraction of a neuron are capable of universal computation. Given that individual neurons can generate complex temporal patterns of activity all by themselves, the range of capabilities possible for even small groups of interconnected neurons are beyond current understanding.

Historically, for many years the predominant view of the function of the nervous system was as a stimulus-response associator. In this conception, neural processing begins with stimuli that activate sensory neurons, producing signals that propagate through chains of connections in the spinal cord and brain, giving rise eventually to activation of motor neurons and thereby to muscle contraction, i.e., to overt responses.

Descartes believed that all of the behaviours of animals, and most of the behaviours of humans, could be explained in terms of stimulus-response circuits, although he also believed that higher cognitive functions such as language were not capable of being explained mechanistically. Charles Sherrington, in his influential 1906 book *The Integrative Action of the Nervous System*, developed the concept of stimulus-response mechanisms in much more detail, and Behaviorism, the school of thought that dominated Psychology through the middle of the 20th century, attempted to explain every aspect of human behaviour in stimulus-response terms.

However, experimental studies of electrophysiology, beginning in the early 20th century and reaching high productivity by the 1940s, showed that the nervous system contains many mechanisms for generating patterns of activity intrinsically, without requiring an external stimulus. Neurons were found to be capable of producing regular sequences of action potentials, or sequences of bursts, even in complete isolation. When intrinsically active neurons are connected to each other in complex circuits, the possibilities for generating intricate temporal patterns become far more extensive. A modern conception views the function of the nervous system partly in terms of stimulus-response chains, and partly in terms of intrinsically generated activity patterns—both types of activity interact with each other to generate the full repertoire of behaviour.

Reflexes and other Stimulus-response Circuits

The simplest type of neural circuit is a reflex arc, which begins with a sensory input and ends with a motor output, passing through a sequence of neurons in between. For example, consider the "withdrawal reflex" causing the hand to jerk back after a hot stove is touched. The circuit begins with sensory receptors in the skin that are activated by harmful levels of heat: a special type of molecular structure embedded in the membrane causes heat to generate an electrical field across the membrane. If the electrical potential change is large enough, it evokes an action potential, which is transmitted along the axon of the receptor cell, into the spinal cord. There the axon makes excitatory synaptic contacts with other cells, some of which project to the same region of the spinal cord, others projecting into the brain. One target is a set of spinal interneurons that project to motor neurons controlling the arm muscles. The interneurons excite the motor neurons, and if the excitation is strong enough, some of the motor neurons generate action potentials, which travel down their axons to the point where they make excitatory synaptic contacts with muscle cells. The excitatory signals induce contraction of the muscle cells, which causes the joint angles in the arm to change, pulling the arm away.

In reality, this straight-forward schema is subject to numerous complications. Although for the simplest reflexes there are short neural paths from sensory neuron to motor neuron, there are also other nearby neurons that participate in the circuit and modulate the response. Furthermore, there are projections from the brain to the spinal cord that are capable of enhancing or inhibiting the reflex.

Although the simplest reflexes may be mediated by circuits lying entirely within the spinal cord, more complex responses rely on signal processing in the brain. Consider, for example, what happens when an object in the periphery of the visual field moves, and a person looks toward it. The initial sensory response, in the retina of the eye, and the final motor response, in the oculomotor nuclei of the brain stem, are not all that different from those in a simple reflex, but the intermediate stages are completely different. Instead of a one or two step chain of processing, the visual signals pass through perhaps a dozen stages of integration, involving the thalamus, cerebral cortex, basal ganglia, superior colliculus, cerebellum, and several brainstem nuclei. These areas perform signal-processing functions that include feature detection, perceptual analysis, memory recall, decision-making, and motor planning.

Feature detection is the ability to extract biologically relevant information from combinations of sensory signals. In the visual system, for example, sensory receptors in the retina of the eye are only individually capable of detecting "points of light" in the outside world. Second-level visual neurons receive input from groups of primary receptors, higher-level neurons receive input from groups of second-level neurons, and so on, forming a hierarchy of processing stages. At each stage, important information is extracted from the signal ensemble and unimportant information is discarded. By the end of the process, input signals representing "points of light" have been transformed into a neural representation of objects in the surrounding world and their properties. The most sophisticated sensory processing occurs inside the brain, but complex feature extraction also takes place in the spinal cord and in peripheral sensory organs such as the retina.

Metabolism

Metabolism is the set of chemical reactions that happen in living organisms to maintain life. These processes allow organisms to grow and reproduce, maintain their structures, and respond to their environments. Metabolism is usually divided into two categories. Catabolism breaks down organic matter, for example to harvest energy in cellular respiration. Anabolism uses energy to construct components of cells such as proteins and nucleic acids. The chemical reactions of metabolism are organized into metabolic pathways, in which one chemical is transformed through a series of steps into another chemical, by a sequence of enzymes.

Enzymes are crucial to metabolism because they allow organisms to drive desirable reactions that require energy and will not occur by themselves, by coupling them to spontaneous reactions that release energy. As enzymes act as catalysts they allow these reactions to proceed quickly and efficiently. Enzymes also allow the regulation of metabolic pathways in response to changes in the cell's environment or signals from other cells.

The metabolism of an organism determines which substances it will find nutritious and which it will find poisonous. For example, some prokaryotes use hydrogen sulfide as a nutrient, yet this gas is poisonous to animals. The speed of metabolism, the metabolic rate, also influences how much food an organism will require. A striking feature of metabolism is the similarity of the basic metabolic pathways and components between even vastly different species. For example, the set of carboxylic acids that are best known as the intermediates in the citric acid cycle are present in all organisms, being found in species as diverse as the unicellular bacteria *Escherichia coli* and huge multicellular organisms like elephants. These striking similarities in metabolism are probably due to the high efficiency of these pathways, and their early appearance in evolutionary history.

Key Biochemicals

Most of the structures that make up animals, plants and microbes are made from three basic classes of molecule: amino acids, carbohydrates and lipids (often called fats). As these molecules are vital for life, metabolic reactions focus on making these molecules during the construction of cells and tissues, or breaking them down and using them as a source of energy, in the digestion and use of food. Many important biochemicals can be joined together to make polymers such as DNA and proteins.

Amino Acids and Proteins

Proteins are made of amino acids arranged in a linear chain and joined together by peptide bonds. Many proteins are the enzymes that catalyse the chemical reactions in metabolism. Other proteins have structural or mechanical functions, such as the proteins that form the cytoskeleton, a system of scaffolding that maintains the cell shape. Proteins are also important in cell signalling, immune responses, cell adhesion, active transport across membranes, and the cell cycle.

Lipids

Lipids are the most diverse group of biochemicals. Their main structural uses are as part of biological membranes such as the cell membrane, or as a source of energy. Lipids are usually defined as hydrophobic or amphipathic biological molecules that will dissolve in organic solvents such as benzene or chloroform. The fats are a large group of compounds that contain fatty acids and glycerol; a glycerol molecule attached to three fatty acid esters is a triacylglyceride. Several variations on this basic structure exist, including alternate backbones such as sphingosine in the sphingolipids, and hydrophilic groups such as phosphate in phospholipids. Steroids such as cholesterol are another major class of lipids that are made in cells.

Carbohydrates

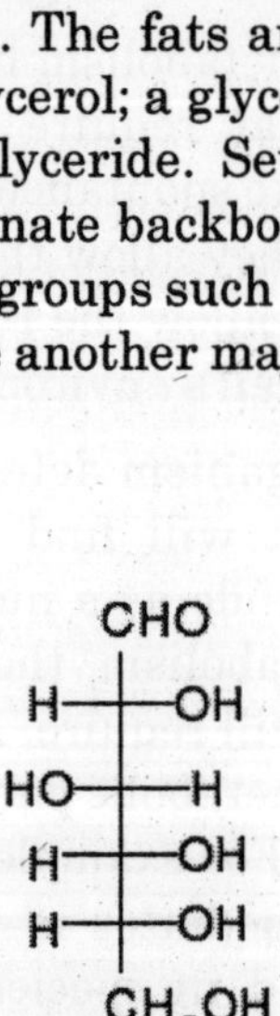

Carbohydrates are straight-chain aldehydes or ketones with many hydroxyl groups that can exist as straight chains or rings. Carbohydrates are the most abundant biological molecules, and fill numerous roles, such as the storage and transport of energy (starch, glycogen) and structural components (cellulose in plants, chitin in animals). The basic carbohydrate units are called monosaccharides and include galactose, fructose, and most importantly glucose. Monosaccharides can be linked together to form polysaccharides in almost limitless ways.

Nucleotides

The two nucleic acids, DNA and RNA are polymers of nucleotides, each nucleotide comprising a phosphate group, a ribose sugar group, and a nitrogenous base. Nucleic acids are critical for the storage and use of genetic information, through the processes of transcription and protein biosynthesis. This information is protected by DNA repair mechanisms and propagated through DNA replication. Many viruses have an RNA genome, for example HIV, which uses reverse transcription to create a DNA template from its viral RNA genome. RNA in ribozymes such as spliceosomes and ribosomes is similar to enzymes as it can catalyse chemical reactions. Individual nucleosides are made by attaching a nucleobase to a ribose sugar. These bases are heterocyclic rings containing nitrogen, classified as purines or pyrimidines. Nucleotides also act as coenzymes in metabolic group transfer reactions.

Coenzymes

Metabolism involves a vast array of chemical reactions, but most fall under a few basic types of reactions that involve the transfer of functional groups. This common chemistry allows cells to use a small set of metabolic intermediates to carry chemical groups between different reactions.

These group-transfer intermediates are called coenzymes. Each class of group-transfer reaction is carried out by a particular coenzyme, which is the substrate for a set of enzymes that produce it, and a set of enzymes that consume it. These coenzymes are therefore continuously being made, consumed and then recycled. One central coenzyme is adenosine triphosphate (ATP), the universal energy currency of cells. This nucleotide is used to transfer chemical energy between different chemical reactions. There is only a small amount of ATP in cells, but as it is continuously regenerated, the human body can use about its own weight in ATP per day.

ATP acts as a bridge between catabolism and anabolism, with catabolic reactions generating ATP and anabolic reactions consuming it. It also serves as a carrier of phosphate groups in phosphorylation reactions.

A vitamin is an organic compound needed in small quantities that cannot be made in the cells. In human nutrition, most vitamins function as coenzymes after modification; for example, all water-soluble vitamins are phosphorylated or are coupled to nucleotides when they are used in cells.

Nicotinamide adenine dinucleotide (NADH), a derivative of vitamin B_3 (niacin), is an important coenzyme that acts as a hydrogen acceptor. Hundreds of separate types of dehydrogenases remove electrons from their substrates and reduce NAD^+ into NADH. This reduced form of the coenzyme is then a substrate for any of the reductases in the cell that need to reduce their substrates. Nicotinamide adenine dinucleotide exists in two related forms in the cell, NADH and NADPH. The NAD^+/NADH form is more important in catabolic reactions, while $NADP^+$/NADPH is used in anabolic reactions.

Minerals and Cofactors

Inorganic elements play critical roles in metabolism; some are abundant (e.g. sodium and potassium) while others function at minute concentrations. About 99% of mammals' mass are the elements carbon, nitrogen, calcium, sodium, chlorine, potassium, hydrogen, phosphorus, oxygen and sulphur. The organic compounds (proteins, lipids and carbohydrates) contain the majority of the carbon and nitrogen and most of the oxygen and hydrogen is present as water.

The abundant inorganic elements act as ionic electrolytes. The most important ions are sodium, potassium, calcium, magnesium, chloride, phosphate, and the organic ion bicarbonate. The maintenance of precise gradients across cell membranes maintains osmotic pressure and pH. Ions are also critical for nerves and muscles, as action potentials in these tissues are produced by the exchange of electrolytes between the extracellular fluid and the cytosol. Electrolytes enter and leave cells through proteins in the cell membrane called ion channels. For example, muscle contraction depends upon the movement of calcium, sodium and potassium through ion channels in the cell membrane and T-tubules.

The transition metals are usually present as trace elements in organisms, with zinc and iron being most abundant. These metals are used in some proteins as cofactors and are essential for the activity of enzymes such as catalase and oxygen-carrier proteins such as hemoglobin. These cofactors are bound tightly to a specific protein; although enzyme cofactors can be modified during catalysis, cofactors always return to their original state after catalysis has taken place. The metal micronutrients are taken up into organisms by specific transporters and bound to storage proteins such as ferritin or metallothionein when not being used.

Catabolism

Catabolism is the set of metabolic processes that break down large molecules. These include breaking down and oxidizing food molecules. The purpose of the catabolic reactions is to provide the energy and components needed by anabolic reactions. The exact nature of these catabolic reactions differ from organism to organism and organisms can be classified based on their sources of energy and carbon (their primary nutritional groups). Organic molecules being used as a source of energy in organotrophs, while lithotrophs use inorganic substrates and phototrophs capture sunlight as chemical energy. However, all these different forms of metabolism depend on redox reactions that involve the transfer of electrons from reduced donor molecules such as organic molecules, water, ammonia, hydrogen sulfide or ferrous ions to acceptor molecules such as oxygen, nitrate or sulfate. In animals these reactions involve complex organic molecules being broken down to simpler molecules, such as carbon dioxide and water. In photosynthetic organisms such as plants and cyanobacteria, these electron-transfer reactions do not release energy, but are used as a way of storing energy absorbed from sunlight.

The most common set of catabolic reactions in animals can be separated into three main stages. In the first, large organic molecules such as proteins, polysaccharides or lipids are digested into their smaller components outside cells. Next, these smaller molecules are taken up by cells and converted to yet smaller molecules, usually acetyl coenzyme A (acetyl-CoA), which releases some energy. Finally, the acetyl group on the CoA is oxidised to water and carbon dioxide in the citric acid cycle and electron transport chain, releasing the energy that is stored by reducing the coenzyme nicotinamide adenine dinucleotide (NAD^+) into NADH.

Digestion

Macromolecules such as starch, cellulose or proteins cannot be rapidly taken up by cells and need to be broken into their smaller units before they can be used in cell metabolism. Several common classes of enzymes digest these polymers. These digestive enzymes include proteases that digest proteins into amino acids, as well as glycoside hydrolases that digest polysaccharides into monosaccharides.

Microbes simply secrete digestive enzymes into their surroundings, while animals only secrete these enzymes from specialized cells in their guts. The amino acids or sugars released by these extracellular enzymes are then pumped into cells by specific active transport proteins.

Energy from Organic Compounds

Further information: Cellular respiration, fermentation, carbohydrate catabolism, fat catabolism and protein catabolism

Carbohydrate catabolism is the breakdown of carbohydrates into smaller units. Carbohydrates are usually taken into cells once they have been digested into monosaccharides. Once inside, the major route of breakdown is glycolysis, where sugars such as glucose and fructose are converted into pyruvate and some ATP is generated.

Pyruvate is an intermediate in several metabolic pathways, but the majority is converted to acetyl-CoA and fed into the citric acid cycle. Although some more ATP is generated in the citric acid cycle, the most important product is NADH, which is made from NAD^+ as the acetyl-CoA is oxidized. This oxidation releases carbon dioxide as a waste product. In anaerobic conditions, glycolysis produces lactate, through the enzyme lactate dehydrogenase re-oxidizing NADH to NAD+ for re-use in glycolysis. An alternative route for glucose breakdown is the pentose phosphate pathway, which reduces the coenzyme NADPH and produces pentose sugars such as ribose, the sugar component of nucleic acids.

Fats are catabolised by hydrolysis to free fatty acids and glycerol. The glycerol enters glycolysis and the fatty acids are broken down by beta oxidation to release acetyl-CoA, which then is fed into the citric acid cycle. Fatty acids release more energy upon oxidation than carbohydrates because carbohydrates contain more oxygen in their structures.

Amino acids are either used to synthesize proteins and other biomolecules, or oxidized to urea and carbon dioxide as a source of energy. The oxidation pathway starts with the removal of the amino group by a transaminase. The amino group is fed into the urea cycle, leaving a deaminated carbon skeleton in the form of a keto acid. Several of these keto acids are intermediates in the citric acid cycle, for example the deamination of glutamate forms α-ketoglutarate. The glucogenic amino acids can also be converted into glucose, through gluconeogenesis.

Energy Transformations

Oxidative Phosphorylation

In oxidative phosphorylation, the electrons removed from food molecules in pathways such as the citric acid cycle are transferred to oxygen and the energy released is used to make ATP. This is done in eukaryotes by a series of proteins in the membranes of mitochondria called the electron transport chain. In prokaryotes, these proteins are found in the cell's inner membrane. These proteins use the energy released from passing electrons from reduced molecules like NADH onto oxygen to pump protons across a membrane.

Pumping protons out of the mitochondria creates a proton concentration difference across the membrane and generates an electrochemical gradient.

This force drives protons back into the mitochondrion through the base of an enzyme called ATP synthase. The flow of protons makes the stalk subunit rotate, causing the active site of the synthase domain to change shape and phosphorylate adenosine diphosphate -turning it into ATP.

Energy from Inorganic Compounds

Chemolithotrophy is a type of metabolism found in prokaryotes where energy is obtained from the oxidation of inorganic compounds. These organisms can use hydrogen, reduced sulphur compounds (such as sulfide, hydrogen sulfide and thiosulfate), ferrous iron (FeII) or ammonia as sources of reducing power and they gain energy from the oxidation of these compounds with electron acceptors such as oxygen or nitrite. These microbial processes are important in global biogeochemical cycles such as acetogenesis, nitrification and denitrification and are critical for soil fertility.

Energy from Light

The energy in sunlight is captured by plants, cyanobacteria, purple bacteria, green sulphur bacteria and some protists. This process is often coupled to the conversion of carbon dioxide into organic compounds, as part of photosynthesis. The energy capture and carbon fixation systems can however operate separately in prokaryotes, as purple bacteria and green sulphur bacteria can use sunlight as a source of energy, while switching between carbon fixation and the fermentation of organic compounds.

In many organisms the capture of solar energy is similar in principle to oxidative phosphorylation, as it involves energy being stored as a proton concentration gradient and this proton motive force then driving ATP synthesis. The electrons needed to drive this electron transport chain come from light-gathering proteins called photosynthetic reaction centres or rhodopsins. Reaction centres are classed into two types depending on the type of photosynthetic pigment present, with most photosynthetic bacteria only having one type, while plants and cyanobacteria have two.

In plants, algae, and cyanobacteria, photosystem II uses light energy to remove electrons from water, releasing oxygen as a waste product. The electrons then flow to the cytochrome b6f complex, which uses their energy to pump protons across the thylakoid membrane in the chloroplast. These protons move back through the membrane as they drive the ATP synthase, as before. The electrons then flow through photosystem I and can then either be used to reduce the coenzyme $NADP^+$.

Anabolism

Anabolism is the set of constructive metabolic processes where the energy released by catabolism is used to synthesize complex molecules.

In general, the complex molecules that make up cellular structures are constructed step-by-step from small and simple precursors. Anabolism involves three basic stages. Firstly, the production of precursors such as amino acids, monosaccharides, isoprenoids and nucleotides, secondly, their activation into reactive forms using energy from ATP, and thirdly, the assembly of these precursors into complex molecules such as proteins, polysaccharides, lipids and nucleic acids.

Organisms differ in how many of the molecules in their cells they can construct for themselves. Autotrophs such as plants can construct the complex organic molecules in cells such as polysaccharides and proteins from simple molecules like carbon dioxide and water. Heterotrophs, on the other hand, require a source of more complex substances, such as monosaccharides and amino acids, to produce these complex molecules. Organisms can be further classified by ultimate source of their energy: photoautotrophs and photoheterotrophs obtain energy from light, whereas chemoautotrophs and chemoheterotrophs obtain energy from inorganic oxidation reactions.

Carbon Fixation

Photosynthesis is the synthesis of carbohydrates from sunlight and carbon dioxide (CO_2). In plants, cyanobacteria and algae, oxygenic photosynthesis splits water, with oxygen produced as a waste product. This process uses the ATP and NADPH produced by the photosynthetic reaction centres, as described above, to convert CO_2 into glycerate 3-phosphate, which can then be converted into glucose. This carbon-fixation reaction is carried out by the enzyme RuBisCO as part of the Calvin – Benson cycle. Three types of photosynthesis occur in plants, C3 carbon fixation, C4 carbon fixation and CAM photosynthesis. These differ by the route that carbon dioxide takes to the Calvin cycle, with C3 plants fixing CO_2 directly, while C4 and CAM photosynthesis incorporate the CO_2 into other compounds first, as adaptations to deal with intense sunlight and dry conditions. In photosynthetic prokaryotes the mechanisms of carbon fixation are more diverse. Here, carbon dioxide can be fixed by the Calvin – Benson cycle, a reversed citric acid cycle, or the carboxylation of acetyl-CoA. Prokaryotic chemoautotrophs also fix CO_2 through the Calvin – Benson cycle, but use energy from inorganic compounds to drive the reaction.

Carbohydrates and Glycans

In carbohydrate anabolism, simple organic acids can be converted into monosaccharides such as glucose and then used to assemble polysaccharides such as starch. The generation of glucose from compounds like pyruvate, lactate, glycerol, glycerate 3-phosphate and amino acids is called gluconeogenesis. Gluconeogenesis converts pyruvate to glucose-6-phosphate through a series of intermediates, many of which are shared with glycolysis. However, this pathway is not simply glycolysis run in reverse, as several steps are catalysed by non-glycolytic enzymes. This is important as it allows the formation and breakdown of glucose to be regulated separately and prevents both pathways from running simultaneously in a futile cycle.

Although fat is a common way of storing energy, in vertebrates such as humans the fatty acids in these stores cannot be converted to glucose through gluconeogenesis as these organisms cannot convert acetyl-CoA into pyruvate; plants do, but animals do not, have the necessary enzymatic machinery. As a result, after long-term starvation, vertebrates need to produce ketone bodies from fatty acids to replace glucose in tissues such as the brain that cannot metabolize fatty acids. In other organisms such as plants and bacteria, this metabolic problem is solved using the glyoxylate cycle, which bypasses the decarboxylation step in the citric acid cycle and allows the transformation of acetyl-CoA to oxaloacetate, where it can be used for the production of glucose.

Polysaccharides and glycans are made by the sequential addition of monosaccharides by glycosyltransferase from a reactive sugar-phosphate donor such as uridine diphosphate glucose (UDP-glucose) to an acceptor hydroxyl group on the growing polysaccharide. As any of the hydroxyl groups on the ring of the substrate can be acceptors, the polysaccharides produced can have straight or branched structures. The polysaccharides produced can have structural or metabolic functions themselves, or be transferred to lipids and proteins by enzymes called oligosaccharyltransferases.

Fatty Acids, Isoprenoids and Steroids

Fatty acids are made by fatty acid synthases that polymerize and then reduce acetyl-CoA units. The acyl chains in the fatty acids are extended by a cycle of reactions that add the actyl group, reduce it to an alcohol, dehydrate it to an alkene group and then reduce it again to an alkane group. The enzymes of fatty acid biosynthesis are divided into two groups, in animals and fungi all these fatty acid synthase reactions are carried out by a single multifunctional type I protein, while in plant plastids and bacteria separate type II enzymes perform each step in the pathway.

Terpenes and isoprenoids are a large class of lipids that include the carotenoids and form the largest class of plant natural products. These compounds are made by the assembly and modification of isoprene units donated from the reactive precursors isopentenyl pyrophosphate and dimethylallyl pyrophosphate. These precursors can be made in different ways. In animals and archaea, the mevalonate pathway produces these compounds from acetyl-CoA, while in plants and bacteria the non-mevalonate pathway uses pyruvate and glyceraldehyde 3-phosphate as substrates. One important reaction that uses these activated isoprene donors is steroid biosynthesis. Here, the isoprene units are joined together to make squalene and then folded up and formed into a set of rings to make lanosterol. Lanosterol can then be converted into other steroids such as cholesterol and ergosterol.

Proteins

Organisms vary in their ability to synthesize the 20 common amino acids. Most bacteria and plants can synthesize all twenty, but mammals can synthesize only eleven nonessential amino acids. Thus, nine essential amino acids must be obtained from food. All amino acids are synthesized from intermediates in glycolysis, the citric acid cycle, or the pentose phosphate pathway. Nitrogen is provided by glutamate and glutamine. Amino acid synthesis depends on the formation of the appropriate alpha-keto acid, which is then transaminated to form an amino acid. Amino acids are made into proteins by being joined together in a chain by peptide bonds. Each different protein has a unique sequence of amino acid residues: this is its primary structure. Just as the letters of the alphabet can be combined to form an almost endless variety of words, amino acids can be linked in varying sequences to form a huge variety of proteins. Proteins are made from amino acids that have been activated by attachment to a transfer RNA molecule through an ester bond. This aminoacyl-tRNA precursor is produced in an ATP-dependent reaction carried out by an aminoacyl tRNA synthetase. This aminoacyl-tRNA is then a substrate for the ribosome, which joins the amino acid onto the elongating protein chain, using the sequence information in a messenger RNA.

Nucleotide Synthesis and Salvage

Nucleotides are made from amino acids, carbon dioxide and formic acid in pathways that require large amounts of metabolic energy. Consequently, most organisms have efficient systems to salvage preformed nucleotides. Purines are synthesized as nucleosides (bases attached to ribose). Both adenine and guanine are made from the precursor nucleoside inosine monophosphate, which is synthesized using atoms from the amino acids glycine, glutamine, and aspartic acid, as well as formate transferred from the coenzyme tetrahydrofolate. Pyrimidines, on the other hand, are synthesized from the base orotate, which is formed from glutamine and aspartate.

Xenobiotics and Redox Metabolism

All organisms are constantly exposed to compounds that they cannot use as foods and would be harmful if they accumulated in cells, as they have no metabolic function. These potentially damaging compounds are called xenobiotics. Xenobiotics such as synthetic drugs, natural poisons and antibiotics are detoxified by a set of xenobiotic-metabolizing enzymes. In humans, these include cytochrome P450 oxidases, UDP-glucuronosyltransferases, and glutathione *S*-transferases. This system of enzymes acts in three stages to firstly oxidize the xenobiotic (phase I) and then conjugate water-soluble groups onto the molecule (phase II). The modified water-soluble xenobiotic can then be pumped out of cells and in multicellular organisms may be further metabolized before being excreted (phase III). In ecology, these reactions are particularly important in microbial biodegradation of pollutants and the bioremediation of contaminated land and oil spills. Many of these microbial reactions are shared with multicellular organisms, but due to the incredible diversity of types of microbes these organisms are able to deal with a far wider range of xenobiotics than multicellular organisms, and can degrade even persistent organic pollutants such as organochloride compounds.

A related problem for aerobic organisms is oxidative stress. Here, processes including oxidative phosphorylation and the formation of disulfide bonds during protein folding produce reactive oxygen species such as hydrogen peroxide. These damaging oxidants are removed by antioxidant metabolites such as glutathione and enzymes such as catalases and peroxidases.

Thermodynamics of Living Organisms

Living organisms must obey the laws of thermodynamics, which describe the transfer of heat and work. The second law of thermodynamics states that in any closed system, the amount of entropy (disorder) will tend to increase. Although living organisms' amazing complexity appears to contradict this law, life is possible as all organisms are open systems that exchange matter and energy with their surroundings. Thus living systems are not in equilibrium, but instead are dissipative systems that maintain their state of high complexity by causing a larger increase in the entropy of their environments. The metabolism of a cell achieves this by coupling the spontaneous processes of catabolism to the non-spontaneous processes of anabolism. In thermodynamic terms, metabolism maintains order by creating disorder.

Regulation and Control

Further information: Metabolic pathway, metabolic control analysis, hormone, regulatory enzymes, and cell signalling.

As the environments of most organisms are constantly changing, the reactions of metabolism must be finely regulated to maintain a constant set of conditions within cells, a condition called homeostasis. Metabolic regulation also allows organisms to respond to signals and interact actively with their environments. Two closely linked concepts are important for understanding how metabolic pathways are controlled. Firstly, the *regulation* of an enzyme in a pathway is how its activity is increased and decreased in response to signals. Secondly, the *control* exerted by this enzyme is the effect that these changes in its activity have on the overall rate of the pathway (the flux through the pathway). For example, an enzyme may show large changes in activity (*i.e.* it is highly regulated) but if these changes have little effect on the flux of a metabolic pathway, then this enzyme is not involved in the control of the pathway.

There are multiple levels of metabolic regulation. In intrinsic regulation, the metabolic pathway self-regulates to respond to changes in the levels of substrates or products; for example, a decrease in the amount of product can increase the flux through the pathway to compensate. This type of regulation often involves allosteric regulation of the activities of multiple enzymes in the pathway. Extrinsic control involves a cell in a multicellular organism changing its metabolism in response to signals from other cells. These signals are usually in the form of soluble messengers such as hormones and growth factors and are detected by specific receptors on the cell surface. These signals are then transmitted inside the cell by second messenger systems that often involved the phosphorylation of proteins.

A very well understood example of extrinsic control is the regulation of glucose metabolism by the hormone insulin. Insulin is produced in response to rises in blood glucose levels. Binding of the hormone to insulin receptors on cells then activates a cascade of protein kinases that cause the cells to take up glucose and convert it into storage molecules such as fatty acids and glycogen.

The metabolism of glycogen is controlled by activity of phosphorylase, the enzyme that breaks down glycogen, and glycogen synthase, the enzyme that makes it. These enzymes are regulated in a reciprocal fashion, with phosphorylation inhibiting glycogen synthase, but activating phosphorylase. Insulin causes glycogen synthesis by activating protein phosphatases and producing a decrease in the phosphorylation of these enzymes.

Evolution

The central pathways of metabolism described above, such as glycolysis and the citric acid cycle, are present in all three domains of living things and were present in the last universal ancestor. This universal ancestral cell was prokaryotic and probably a methanogen that had extensive amino acid, nucleotide, carbohydrate and lipid metabolism. The retention of these ancient pathways during later evolution may be the result of these reactions being an optimal solution to their particular metabolic problems, with pathways such as glycolysis and the citric acid cycle producing their end products highly efficiently and in a minimal number of steps. The first pathways of enzyme-based metabolism may have been parts of purine nucleotide metabolism, with previous metabolic pathways being part of the ancient RNA world. Many models have been proposed to describe the mechanisms by which novel metabolic pathways evolve. These include the sequential addition of novel enzymes to a short ancestral pathway, the duplication and then divergence of entire pathways as well as the recruitment of pre-existing enzymes and their assembly into a novel reaction pathway. The relative importance of these mechanisms is unclear, but genomic studies have shown that enzymes in a pathway are likely to have a shared ancestry, suggesting that many pathways have evolved in a step-by-step fashion with novel functions being created from pre-existing steps in the pathway. An alternative model comes from studies that trace the evolution of proteins' structures in metabolic networks, this has suggested that enzymes are pervasively recruited, borrowing enzymes to perform similar functions in different metabolic pathways (evident in the MANET database) These recruitment processes result in an evolutionary enzymatic mosaic. A third possibility is that some parts of metabolism might exist as "modules" that can be reused in different pathways and perform similar functions on different molecules.

As well as the evolution of new metabolic pathways, evolution can also cause the loss of metabolic functions. For example, in some parasites metabolic processes that are not essential for survival are lost and preformed amino acids, nucleotides and carbohydrates may instead be scavenged from the host. Similar reduced metabolic capabilities are seen in endosymbiotic organisms.

Investigation and Manipulation

Classically, metabolism is studied by a reductionist approach that focuses on a single metabolic pathway. Particularly valuable is the use of radioactive tracers at the whole-organism, tissue and cellular levels, which define the paths from precursors to final products by identifying radioactively labelled intermediates and products. The enzymes that catalyse these chemical reactions can then be purified and their kinetics and responses to inhibitors investigated. A parallel approach is to identify the small molecules in a cell or tissue; the complete set of these molecules is called the metabolome. Overall, these studies give a good view of the structure and function of simple metabolic pathways, but are inadequate when applied to more complex systems such as the metabolism of a complete cell. However, it is now possible to use this genomic data to reconstruct complete networks of biochemical reactions and produce more holistic mathematical models that may explain and predict their behaviour.

These models are especially powerful when used to integrate the pathway and metabolite data obtained through classical methods with data on gene expression from proteomic and DNA microarray studies. Using these techniques, a model of human metabolism has now been produced, which will guide future drug discovery and biochemical research. These models are now being used in network analysis, to classify human diseases into groups that share common proteins or metabolites.

Bacterial metabolic networks seem to be a striking example of bow-tie organization, an architecture able to input a wide range of nutrients and produce a large variety of products and complex macromolecules using a relatively few intermediate common currencies. A major technological application of this information is metabolic engineering. Here, organisms such as yeast, plants or bacteria are genetically modified to make them more useful in biotechnology and aid the production of drugs such as antibiotics or industrial chemicals such as 1,3-propanediol and shikimic acid. These genetic modifications usually aim to reduce the amount of energy used to produce the product, increase yields and reduce the production of wastes.

7

Biological and Synthetic Dioxygen Carriers

Most organisms require molecular oxygen in order to survive. The dioxygen is used in a host of biochemical transformations, although most is consumed in the reaction that is the terminal (or primary) step of oxidative phosphorylation. For some small animals and for plants, where the surface-to-volume ratio is large, an adequate supply of dioxygen can be obtained from simple diffusion across cell membranes. The dioxygen may be extracted from air or water; for plants that produce dioxygen in photosynthesis, it is also available endogenously. For other organisms, particularly those with non-passive lifestyles, from scorpions to whales, diffusion does not supply sufficient dioxygen for respiration.

An elegant three-component system has evolved to transport dioxygen from regions of high abundance-water (at least if free of pollutant reductants) and air-to regions of relatively low abundance and high demand-the interior cells of the organism. The central component is a dioxygen-carrier protein. In the three chemically distinct carriers that have evolved and are found today, the dioxygen-binding site in the protein, that is, the so-called "active site," is a complex either of copper or of iron. For hemoglobins, the most widely distributed family of dioxygen carriers, the active site has long been known to consist of an iron porphyrin (heme) group embedded in the protein. Almost all hemoglobins share the basic structure. Hemocyanin and hemerythrin, the other two biological dioxygen carriers, feature pairs of copper atoms and iron atoms, respectively, at the active sites.

The second component of the dioxygen-transport system facilitates the sequestration of dioxygen by the dioxygen-carrier protein. Specialized organs, such as lungs in air-breathing creatures or gills in fish, offer a very large surface area to the outside environment in order to facilitate diffusion. The third component is the delivery system. The oxygen carrier is dissolved or suspended in a fluid, called blood plasma or hemolymph, that is pumped throughout the animal by another specialized organ, the heart, through a network of tubes, the blood vessels.

In many organisms an additional dioxygen-binding protein, which stores dioxygen, is located in tissues that are subject to sudden and high dioxygen demand, such as muscles. These dioxygen-storage proteins are prefixed *myo*(from the Greek root *mys* for muscle). Thus for the dioxygen-transport protein hemerythrin there exists a chemically similar dioxygen-storage protein myohemerythrin.

For the hemoglobin family the corresponding storage protein is called myoglobin. Interestingly, some organisms that use hemocyanin as the dioxygen*transport* protein use myoglobin as the *dioxygen-storage* protein.

At the centre of biological dioxygen transport are transition-metal complexes of iron or copper. To model such systems, chemists have prepared several synthetic oxygen carriers, especially of iron and cobalt porphyrins. In this chapter the structures and properties of biological and nonbiological oxygen carriers are described, with particular attention to the hemoglobin family.

This family has been studied in more detail than any other group of proteins, and as a result a deeper understanding of the relationships among structure, properties, and biological function (i.e., physiology) exists. The central focus of this chapter is to delineate chemical features that determine the affinity of an active site, especially an iron porphyrin, for molecular oxygen.

In order to develop this theme, macroscopic (thermodynamic and kinetic) factors associated with dioxygen binding and release are summarized first. The nonbiological chemistry of iron and copper in the presence of dioxygen is described briefly to elucidate the key role that the protein plays in supporting oxygen transport by preventing irreversible oxidation of the binding site or of its ligands.

The macroscopic behaviour of the biological systems is related to the microscopic picture that has been developed over the last 30 years from x-ray crystallographic studies and a miscellany of spectroscopic probes of the oxygen-binding site.

Relationships between the geometry and charge distribution in the metal-dioxygen moiety and the nature of the interactions between this moiety and its surroundings are examined. Nonbiological dioxygen carriers have proved particularly useful in providing precise and accurate structural information as well as thermodynamic and kinetic data against which the corresponding data from biological oxygen carriers can be contrasted.

The bioinorganic chemistry of the hemoglobin family of oxygen binders is particularly amenable to study by means of small-molecule model systems: four of the five ligands that make up the active site are provided by a square-planar tetradentate ligand, the protoporphyrin IX dianion.

One axial ligand in hemoglobin, imidazole from a histidine residue, is provided by the protein, and the remaining sixth coordination site is available for the exogenous ligand, e.g., dioxygen or carbon monoxide. Thus a model system that approximates the stereochemistry of the active site in hemoglobin may be assembled from an iron(II) porphyrin and a ligand, such as imidazole or pyridine. On the other hand, in hemocyanin and hemerythrin most of the ligands are supplied by the protein.

Thus the assembly of a model system that provides appropriate ligands correctly disposed around the pair of metal atoms poses a major synthetic challenge, especially for hemocyanin, where details on the number, type, and arrangement of ligands have been difficult to establish. Many aspects of the physical, inorganic, and structural chemistry underlying biological oxygen transport and utilization have been clarified through model systems.

Requirements for Effective Oxygen Carriers

In order for dioxygen transport to be more efficient than simple diffusion through cell membranes and fluids, it is not sufficient that a metalloprotein merely binds dioxygen.

Not only is there an optimal affinity of the carrier for dioxygen, but also, and more importantly, the carrier must bind *and* release dioxygen at a rapid rate.

Oxygen Cycle

The Oxygen cycle is the biogeochemical cycle that describes the movement of oxygen within its three main reservoirs: the atmosphere (air), the total content of biological matter within the biosphere (the global sum of all ecosystems), and the lithosphere (Earth's crust).

Failures in the oxygen cycle within the hydrosphere (the combined mass of water found on, under, and over the surface of a planet) can result in the development of hypoxic zones. The main driving factor of the oxygen cycle is photosynthesis, which is responsible for the modern Earth's atmosphere and life.

Reservoirs

By far the largest reservoir of Earth's oxygen is within the silicate and oxide minerals of the crust and mantle (99.5%). Only a small portion has been released as free oxygen to the biosphere (0.01%) and atmosphere

(0.36%). The main source of atmospheric oxygen is photosynthesis, which produces sugars and oxygen from carbon dioxide and water:-

$$6CO_2 + 6H_2O + energy \rightarrow C_6H_{12}O_6 + 6O_2$$

Photosynthesizing organisms include the plant life of the land areas as well as the phytoplankton of the oceans. The tiny marine cyanobacterium Prochlorococcus was discovered in 1986 and accounts for more than half of the photosynthesis of the open ocean.

An additional source of atmospheric oxygen comes from photolysis, whereby high energy ultraviolet radiation breaks down atmospheric water and nitrous oxide into component atoms. The free H and N atoms escape into space leaving O_2 in the atmosphere:

$$2H_2O + energy \rightarrow 4H + O_2$$

$$2N_2O + energy \rightarrow 4N + O_2$$

The main way oxygen is lost from the atmosphere is via respiration and decay, mechanisms in which animal life and bacteria consume oxygen and release carbon dioxide.

Because lithospheric minerals are oxidised in oxygen, chemical weathering of exposed rocks also consumes oxygen. An example of surface weathering chemistry is formation of iron-oxides (rust):

$$4FeO + O_2 \rightarrow 2Fe_2O_3$$

Oxygen is also cycled between the biosphere and lithosphere. Marine organisms in the biosphere create calcium carbonate shell material ($CaCO_3$) that is rich in oxygen. When the organism dies its shell is deposited on the shallow sea floor and buried over time to create the limestone rock of the lithosphere. Weathering processes initiated by organisms can also free oxygen from the lithosphere. Plants and animals extract nutrient minerals from rocks and release oxygen in the process.

Capacities and Fluxes

The following tables offer estimates of oxygen cycle reservoir capacities and fluxes. These numbers are based primarily on estimates from (Walker, J.C.G.) :

Table 1: Major reservoirs involved in the oxygen cycle

Reservoir	*Capacity (kg O_2)*	*Flux In/Out (kg O_2 per year)*	*Residence Time (years)*
Atmosphere	$1.4 * 10^{18}$	$30{,}000 * 10^{10}$	4,500
Biosphere	$1.6 * 10^{16}$	$30{,}000 * 10^{10}$	50
Lithosphere	$2.9 * 10^{20}$	$60 * 10^{10}$	500,000,000

Table 2: Annual gain and loss of atmospheric oxygen (Units of 10^{10} kg O_2 per year)

Photosynthesis (land)	16,500
Photosynthesis (ocean)	13,500
Photolysis of N_2O	1.3
Photolysis of H_2O	0.03
Total Gains	~ 30,000
Losses- Respiration and Decay	
Aerobic Respiration	23,000
Microbial Oxidation	5,100
Combustion of Fossil Fuel (anthropogenic)	1,200
Photochemical Oxidation	600
Fixation of N_2 by Lightning	12
Fixation of N_2 by Industry (anthropogenic)	10
Oxidation of Volcanic Gases	5
Losses- Weathering	
Chemical Weathering	50
Surface Reaction of O_3	12
Total Losses	~ 30,000

Ozone

The presence of atmospheric oxygen has led to the formation of ozone(O_3) and the ozone layer within the stratosphere. The ozone layer is extremely important to modern life as it absorbs harmful ultraviolet radiation:

$$O_2 + \text{uv energy} \rightarrow 2O$$

$$O + O_2 \rightarrow O_3$$

Apparent Oxygen Utilisation

In freshwater or marine systems apparent oxygen utilisation (AOU) is the difference between the measured dissolved oxygen concentration and its equilibrium saturation concentration in water with the same physical and chemical properties. Such differences typically occur when biological activity acts to change the ambient concentration of oxygen. For example, primary production liberates oxygen and increases its concentration, while respiration consumes it and decreases its concentration.

Consequently, the AOU of a water sample represents the sum of the biological activity that the sample has experienced since it was last in equilibrium with the atmosphere. In shallow water systems (e.g. lakes), the full water column is generally in close contact with the atmosphere, so oxygen concentrations are typically close to saturation. In deep water systems (e.g. oceans), water can be out of contact with the atmosphere for extremely long periods of time (years, decades, centuries) and large AOU values are possible.

Oxygen–haemoglobin Dissociation Curve

The oxygen–haemoglobin dissociation curve (or oxygen–hemoglobin dissociation curve) plots the proportion of haemoglobin in its saturated form on the vertical axis against the prevailing oxygen tension on the horizontal axis. The oxyhaemoglobin dissociation curve is an important tool for understanding how our blood carries and releases oxygen. Specifically, the oxyhaemoglobin dissociation curve relates oxygen saturation (SO_2) and partial pressure of oxygen in the blood (PO_2), and is determined by what is called "haemoglobin's affinity for oxygen"; that is, how readily haemoglobin acquires and releases oxygen molecules into the fluid that surrounds it.

Background

Haemoglobin (Hb), a globular protein, is the primary vehicle for transporting oxygen in the blood. Oxygen is also carried dissolved in the blood's plasma, but to a much lesser degree. Haemoglobin is contained in erythrocytes, more commonly referred to as red blood cells. Under certain conditions, oxygen bound to the haemoglobin is released into the blood's plasma and absorbed into the tissues. Each haemoglobin molecule has the capacity to carry four oxygen molecules. How much of that capacity is filled by oxygen at any time is called the oxygen saturation.

Expressed as a percentage, the oxygen saturation is the ratio of the amount of oxygen bound to the haemoglobin, to the oxygen-carrying capacity of the haemoglobin. The oxygen-carrying capacity of haemoglobin is determined by the type of haemoglobin present in the blood. The amount of oxygen bound to the haemoglobin at any time is related, in large part, to the partial pressure of oxygen to which the haemoglobin is exposed.

In the lungs, at the alveolar–capillary interface, the partial pressure of oxygen is typically high, and therefore the oxygen binds readily to haemoglobin that is present. As the blood circulates to other body tissue in which the partial pressure of oxygen is less, the haemoglobin releases the oxygen into the tissue because the haemoglobin cannot maintain its full bound capacity of oxygen in the presence of lower oxygen partial pressures.

Sigmoidal Shape

It is usually a sigmoid plot. A haemoglobin molecule can bind up to four oxygen molecules in a reversible way.

The shape of the curve results from the interaction of bound oxygen molecules with incoming molecules. The binding of the first molecule is difficult. However, this facilitates the binding of the second and third molecules, and it is only when the fourth molecule is to be bound that the difficulty increases, partly as a result of crowding of the haemoglobin molecule, partly as a natural tendency of oxygen to dissociate.

In its most simple form, the oxyhaemoglobin dissociation curve describes the relation between the partial pressure of oxygen (x axis) and the oxygen saturation (y axis). Haemoglobin's affinity for oxygen increases as successive molecules of oxygen bind. More molecules bind as the oxygen partial pressure increases until the maximum amount that can be bound is reached. As this limit is approached, very little additional binding occurs and the curve levels out as the haemoglobin becomes saturated with oxygen. Hence the curve has a sigmoidal or S-shape. At pressures above about 60 mmHg, the standard dissociation curve is relatively flat, which means that the oxygen content of the blood does not change significantly even with large increases in the oxygen partial pressure. To get more oxygen to the tissue would require blood transfusions to increase the haemoglobin count (and hence the oxygen-carrying capacity), or supplemental oxygen that would increase the oxygen dissolved in plasma.

Although binding of oxygen to haemoglobin continues to some extent for pressures about 50 mmHg, as oxygen partial pressures decrease in this steep area of the curve, the oxygen is unloaded to peripheral tissue readily as the haemoglobin's affinity diminishes.

The partial pressure of oxygen in the blood at which the haemoglobin is 50% saturated, typically about 26.6 mmHg for a healthy person, is known as the P50. The P50 is a conventional measure of haemoglobin affinity for oxygen. In the presence of disease or other conditions that change the haemoglobin's oxygen affinity and, consequently, shift the curve to the right or left, the P50 changes accordingly. An increased P50 indicates a rightward shift of the standard curve, which means that a larger partial pressure is necessary to maintain a 50% oxygen saturation. This indicates a decreased affinity. Conversely, a lower P50 indicates a leftward shift and a higher affinity.

The 'plateau' portion of the oxyhaemoglobin dissociation curve is the range that exists at the pulmonary capillaries (minimal reduction of oxygen transported until the $p(O_2)$ falls 50 mmHg).

The 'steep' portion of the oxyhaemoglobin dissociation curve is the range that exists at the systemic capillaries (a small drop in systemic capillary $p(O_2)$ can result in the release of large amounts of oxygen for the metabolically active cells).

To see the relative affinities of each successive oxygen as you remove/add oxygens from/to the haemoglobin from the curve compare the relative increase/decrease in $p(O_2)$ needed for the corresponding increase/decrease in $s(O_2)$.

Factors Shifting Curve

Many factors influence the affinity of this binding and alter the shape of the curve:

	right shift (low affinity for O_2)	left shift (high affinity for O_2)
temperature	high	low
2.3-DPG	high	low
$p(CO_2)$	high	low
p(CO)	low	high
pH (Bohr effect)	low (acidosis)	high (alkalosis)m
type of haemoglobin	adult haemoglobin	foetal haemoglobin

Left shift of the curve is a sign of haemoglobin's increased affinity for oxygen (e.g. at the lungs). Similarly, right shift shows decreased affinity, as would appear with an increase in body temperature, hydrogen ion, 2,3-diphosphoglycerate (also known as bisphosphoglycerate) or carbon dioxide concentration (the Bohr effect)

Carbon monoxide has a much higher affinity for haemoglobin than oxygen does. In carbon monoxide poisoning, oxygen cannot be transported and released to body tissues thus resulting in hypoxia.

With foetal haemoglobin, the shift facilitates diffusion of oxygen across the placenta. The oxygen dissociation curve for myoglobin exists even further to the left.

The causes of shift to right can be remembered using the mnemonic, "CADET, face Right!" for CO2, Acid, 2,3-DPG, Exercise and Temperature. Factors that move the oxygen dissociation curve to the right are those physiological states where tissues need more oxygen. For example during exercise, muscles have a higher metabolic rate, and consequently need more oxygen, produce more carbon dioxide and lactic acid, and their temperature rises.

Factors that Affect the Standard Dissociation Curve

The strength with which oxygen binds to haemoglobin is affected by several factors. In effect these factors shift or reshape the oxyhaemoglobin curve ("the standard curve"). The standard curve is shifted to the right by an increase in temperature, 2,3-DPG, or pCO_2, or a decrease in pH.

A rightward shift indicates that the haemoglobin under study has a decreased affinity for oxygen. This makes it more difficult for haemoglobin to bind to oxygen (requiring a higher partial pressure of oxygen to achieve the same oxygen saturation), but it makes it easier for the haemoglobin to release oxygen bound to it.

The effect of this rightward shift of the curve increases the partial pressure of oxygen in the tissues when it is most needed, such as during exercise, or haemorrhagic shock.

In contrast, the curve is shifted to the left by the opposite of these conditions. This leftward shift indicates that the haemoglobin under study has an increased affinity for oxygen so that haemoglobin binds oxygen more easily, but unloads it more reluctantly.

Variation of the Hydrogen Ion Concentration

This changes the blood's pH. A decrease in pH shifts the standard curve to the right, while an increase shifts it to the left. This is known as the Bohr effect.

Effects of Carbon Dioxide

Carbon dioxide affects the curve in two ways: first, it influences intracellular pH (the Bohr effect), and second, CO_2 accumulation causes carbamino compounds to be generated through chemical interactions, which bind to haemoglobin forming carbaminohaemoglobin.

Low levels of carbamino compounds have the effect of shifting the curve to the right, while higher levels cause a leftward shift. However, this isn't the overriding effect of CO_2 accumulation. Only about 5–10% of the total CO_2 content of blood is transported as carbamino compounds.

Most of the CO_2 content (80–90%) is transported as bicarbonate ions. The formation of a bicarbonate ion will release a proton into the plasma.

Hence, the elevated CO_2 content creates a respiratory acidosis and shifts the oxygen–haemoglobin dissociation curve to the right.

Effects of 2,3-DPG

2,3-Disphosphoglycerate or 2,3-DPG (also 2,3-bisphosphoglycerate or 2,3-BPG) is an organophosphate, which is created in erythrocytes during glycolysis. The production of 2,3-DPG is likely an important adaptive mechanism, because the production increases for several conditions in the presence of diminished peripheral tissue O_2 availability, such as hypoxaemia, chronic lung disease, anaemia, and congestive heart failure, among others. High levels of 2,3-DPG shift the curve to the right, while low levels of 2,3-DPG cause a leftward shift, seen in states such as septic shock and hypophosphataemia.

Temperature

Temperature does not have such a dramatic effect compared to the previous factors, but hyperthermia causes a rightward shift, while hypothermia causes a leftward shift.

Carbon Monoxide

Haemoglobin binds with carbon monoxide 240 times more readily than with oxygen. The presence of carbon monoxide on one of the 4 haem sites causes the oxygen on the other haem sites to bind with greater affinity.

This makes it difficult for the haemoglobin to release oxygen to the tissues and has the effect of shifting the curve to the left (as well as downward, due to direct competitive effects of carbon monoxide). With an increased level of carbon monoxide, a person can suffer from severe tissue hypoxia while maintaining a normal pO_2.

Effects of Methaemoglobinaemia

Methaemoglobinaemia is a form of abnormal haemoglobin where ferrous (Fe^{2+}), which is normally found in haemoglobin, is converted to the ferric (Fe^{3+}) state. This causes a leftward shift in the curve as methaemoglobin does not unload O2 from Hb. However, methaemoglobin has increased affinity for cyanide, and is therefore useful in the treatment of cyanide poisoning.

Fetal Haemoglobin

Fetal haemoglobin (HbF) is structurally different from normal haemoglobin (Hb). The fetal dissociation curve is shifted to the left relative to the curve for the normal adult. Typically, fetal arterial oxygen pressures are low, and hence the leftward shift enhances the placental uptake of oxygen. At the placenta there is a higher concentration of 2,3-DPG formed. This binds more readily to adult haemoglobin but not to fetal haemoglobin. This causes the adult Hb to release more oxygen at the placenta to be taken up by the fetus. Fetal Hb is made up of gamma chains not beta ones, and 2,3-DPG does not bind readily to gamma chains, hence it does not give up its oxygen.

Dioxygen or Molecular Oxygen

Atoms

The atom is a basic unit of matter consisting of a dense, central nucleus surrounded by a cloud of negatively charged electrons. The atomic nucleus contains a mix of positively charged protons and electrically neutral neutrons (except in the case of hydrogen-1, which is the only stable nuclide with no neutron). The electrons of an atom are bound to the nucleus by the electromagnetic force.

An atom containing an equal number of protons and electrons is electrically neutral, otherwise it has a positive or negative charge and is an ion. An atom is classified according to the number of protons and neutrons in its nucleus: the number of protons determines the chemical element (noted Z), and the number of neutrons determine the isotope of the element. In chemistry, a group of atoms can remain bound to each other, forming a molecule. Over 99.9% of an atom's mass is concentrated in the nucleus (order of magnitude 10^{-15} m), with protons and neutrons having roughly equal mass.

The electrons in an atom are attracted to the protons in the nucleus by the electromagnetic force. This force binds the electrons inside an electrostatic potential well surrounding the smaller nucleus, which means that an external source of energy is needed in order for the electron to escape. The closer an electron is to the nucleus, the greater the attractive force. Hence electrons bound near the centre of the potential well require more energy to escape than those at greater separations.

Electrons that are bound to atoms possess a set of stable energy levels, or orbitals, and can undergo transitions between them by absorbing or emitting photons that match the energy differences between the levels. The electrons determine the chemical properties of an element, and strongly influence an atom's magnetic properties.

Molecules

A molecule is defined as an electrically neutral group of at least two atoms in a definite arrangement held together by very strong (covalent) chemical bonds. Molecules are distinguished from polyatomic ions in this strict sense. In organic chemistry and biochemistry, the term molecule is used less strictly and also is applied to charged organic molecules and biomolecules.

A molecule may consist of atoms of a single chemical element, as with oxygen (O_2), or of different elements, as with water (H_2O). Atoms and complexes connected by non-covalent bonds such as hydrogen bonds or ionic bonds are generally not considered single molecules.

Oxygen Molecule and Covalent Bounding

The common allotrope of elemental oxygen on Earth is called dioxygen, O_2. It has a bond length of 121 pm and a bond energy of 498 kJ mol^{-1}. This is the form that is used by complex forms of life, such as animals, in cellular respiration and is the form that is a major part of the Earth's atmosphere.

Oxygen is the most abundant chemical element, by mass, in our biosphere, air, sea and land. Oxygen is the third most abundant chemical element in the universe, after hydrogen and helium. About 0.9% of the Sun's mass is oxygen. Oxygen constitutes 49.2% of the Earth's crust by mass and is the major component of the world's oceans (88.8% by mass). Oxygen gas is the second most common component of the Earth's atmosphere, taking up 21.0% of its volume and 23.1% of its mass (some 1015 tonnes).

Earth is unusual among the planets of the Solar System in having such a high concentration of oxygen gas in its atmosphere: Mars (with 0.1% O2 by volume) and Venus have far lower concentrations. However, the O_2 surrounding these other planets is produced solely by ultraviolet radiation impacting oxygen-containing molecules such as carbon dioxide.

It is a colourless, odourless and tasteless gas.

The presence of large amounts of dissolved and free oxygen in the oceans and atmosphere may have driven most of the anaerobic organisms then living to extinction during the oxygen catastrophe about 2.4 billion years ago. However, cellular respiration using O_2 enables aerobic organisms to produce much more ATP than anaerobic organisms, helping the former to dominate Earth's biosphere. Photosynthesis and cellular respiration of O_2 allowed for the evolution of eukaryotic cells and ultimately complex multicellular organisms such as plants and animals.

From Atomic to Molecular Oxygen : Covalent Bounding

A covalent bond is a form of chemical bonding that is characterized by the sharing of pairs of electrons between atoms, and other covalent bonds. In short, the attraction-to-repulsion stability that forms between atoms when they share electrons is known as covalent bonding.

The term covalent bond dates from 1939. The prefix co-means jointly, associated in action, partnered to a lesser degree, etc.; thus a "co-valent bond", essentially, means that the atoms share "valence".

Walter Heitler and Fritz London are credited with the first successful quantum mechanical explanation of a chemical bond, specifically that of molecular hydrogen, in 1927. Their work was based on the valence bond model, which assumes that a chemical bond is formed when there is good overlap between the atomic orbitals of participating atoms. These atomic orbitals are known to have specific angular relationships between each other, and thus the valence bond model can successfully predict the bond angles observed in simple molecules.

The basic principle is that the properties of chemical elements (atoms) dependent on their electronic structure including electrons called "outer layers". They are composed of electrons less bounded to the nucleus and occupy "layers" usually incomplete.

This incompleteness is energetically unfavourable and pushes atom to start an electronic boundings (therefore, in fine chemicals) with other atoms. The oxygen has only four electrons in its outer layer but would be six for this layer is full (inner layer of type s with 2 outer shell electrons and p-type which is full with 6 electrons: rare gas configuration).

Electron "Time Shared" on the Bounding Orbitals

The oxygen gas solves this problem by combining two oxygen atoms in a molecule of oxygen (that you breathe in nitrogen mixture at ambient). This molecule is formed by "pooling" of two binding electrons by "time sharing" which means that technically each atom has six from time to time and is happy ! This partial electrons equally pooling on each atom is typical of the covalent bonding.

When two atoms approach each other, it is manifest from a certain distance (about 1 nanometer) an attraction between the nucleus of one and the other electrons (and vice versa). This attraction becomes significant when the distance between the nuclei is more than 500 picometer.

But from this distance the repulsion between charges of same sign tend to compensate for the attraction between opposite sign charges.

An equilibrium set-up at a definite distance between the 2 nuclei, characteristic for each atom. Its value is, by way of examples, from 106 pm for the hydrogen and 121 pm for oxygen.

Parameters of the Dioxygen

The parameters of the dioxygen are the following :

- Nuclei mass $^{16}O = 2.678\times10^{-26}$ kg
  - o 8 neutons mass 1,674 927 28(29)$\times10^{-27}$ kg = 13,399 418 (24)$\times10^{-27}$ kg
  - o 8 protons mass 1,672 621 71(29)$\times10^{-27}$ kg = 13,380 973 (68)$\times10^{-27}$ kg
- Electron mass e = 9.109×10^{-31} kg
- Mass ratio : 3.40×10^{-5}
- Constant force rate : k = 11.77 N/cm
- Interatomic distance (between centrum of each nuclei): 120,74$\times10^{-12}$ m (picomètre = pm)
- Covalent radius : 73×10^{-12} m (pm).

Dioxygen Reactions

In nature, free oxygen is produced by the light-driven splitting of water during oxygenic photosynthesis. Green algae and cyanobacteria in marine environments provide about 70% of the free oxygen produced on earth and the rest is produced by terrestrial plants.

A simplified overall formula for photosynthesis is:

$6\ CO_2 + 6\ H_2O + \text{photons} \rightarrow C_6H_{12}O_6 + 6\ O_2$ (or simply carbon dioxide + water + sunlight → glucose + dioxygen)

Photolytic oxygen evolution occurs in the thylakoid membranes of photosynthetic organisms and requires the energy of four photons. Many steps are involved, but the result is the formation of a proton gradient across the thylakoid membrane, which is used to synthesize ATP via photophosphorylation. The O_2 remaining after oxidation of the water molecule is released into the atmosphere.

Molecular dioxygen, O_2, is essential for cellular respiration in all aerobic organisms. Oxygen is used in mitochondria to help generate adenosine triphosphate (ATP) during oxidative phosphorylation. The reaction for aerobic respiration is essentially the reverse of photosynthesis and is simplified as:

$$C_6H_{12}O_6 + 6\ O_2 \rightarrow 6\ CO_2 + 6\ H_2O + 2880\ \text{kJ mol}^{-1}$$

In vertebrates, O_2 is diffused through membranes in the lungs and into red blood cells. Hemoglobin binds O_2, changing its colour from bluish red to bright red. Other animals use hemocyanin (molluscs and some arthropods) or hemerythrin (spiders and lobsters). A liter of blood can dissolve 200 cm^3 of O_2.

Reactive oxygen species, such as superoxide ion (O^{-2}) and hydrogen peroxide (H_2O_2), are dangerous by-products of oxygen use in organisms. Parts of the immune system of higher organisms, however, create peroxide, superoxide, and singlet oxygen to destroy invading microbes. Reactive oxygen species also play an important role in the hypersensitive response of plants against pathogen attack.

An adult human in rest inhales 1.8 to 2.4 grams of oxygen per minute. This amounts to more than 6 billion tonnes of oxygen inhaled by humanity per year.

Dioxygen in Biological Reactions

Dioxygen (O_2) plays an important role in the energy metabolism of living organisms. Free oxygen is produced in the biosphere through photolysis (light-driven oxidation and splitting) of water during photosynthesis in cyanobacteria, green algae, and plants. During oxidative phosphorylation in cellular respiration, oxygen is reduced to water, thus closing the biological water-oxygen redox cycle.

Photosynthesis

In nature, free oxygen is produced by the light-driven splitting of water during oxygenic photosynthesis. Green algae and cyanobacteria in marine environments provide about 70% of the free oxygen produced on earth. The remainder is produced by terrestrial plants, although for example, almost all oxygen produced in tropical forests is consumed by organisms living there.

A simplified overall formula for photosynthesis is:

$$6CO_2 + 6H_2O + \text{photons} \rightarrow C_6H_{12}O_6 + 6O_2$$

(or simply carbon dioxide + water + sunlight → glucose + oxygen)

Photolytic oxygen evolution during photosynthesis occurs via the light-dependent oxidation of water to molecular oxygen and can be written as the following simplified chemical reaction: $2H_2O \rightarrow 4e^- + 4H^+ + O_2$

The reaction occurs in the thylakoid membranes of cyanobacteria, and algal and plant chloroplasts and requires the energy of four photons. The electrons from the oxidized water molecules replace electrons in the P_{680} component of photosystem II, which have been removed into an electron transport chain via light-dependent excitation and resonance energy transfer onto plastoquinone. Photosytem II therefore has also been referred to as water-plastoquinone oxido-reductase. The protons from the oxidized water molecules are released into the thylakoid lumen, thus contributing to the generation of a proton gradient across the thylakoid membrane. This proton gradient is the driving force for ATP synthesis via photophosphorylation and coupling the absorption of light energy and photolysis of water to the creation of chemical energy during photosynthesis. The O_2 remaining after oxidation of the water molecule is released into the atmosphere.

Water oxidation is catalysed by a manganese-containing enzyme complex known as the oxygen evolving complex (OEC) or water-splitting complex found associated with the lumenal side of thylakoid membranes. Manganese is an important cofactor, and calcium and chloride are also required for the reaction to occur.

Oxygen Uptake and Transport

In vertebrates, oxygen uptake is carried out by the following processes:

Oxygen diffuses through membranes and into red blood cells after inhalation into the lungs. The heme group of hemoglobin binds oxygen when it is present, changing hemoglobin's colour from bluish red to bright red. Vertebrate animals use hemoglobin in their blood to transport oxygen from their lungs to their tissues, but other animals use hemocyanin (molluscs and some arthropods) or hemerythrin (spiders and lobsters). A liter of blood can dissolve 200 cc of oxygen gas, which is much more than water can dissolve.

After being carried in blood to a body tissue in need of oxygen, O_2 is handed-off from the heme group to monooxygenase, an enzyme that also has an active site with an atom of iron. Monooxygenase uses oxygen to catalyse many oxidation reactions in the body. Carbon dioxide, a waste product, is released from the cell and into the blood, where it combines with bicarbonate and hemoglobin for transport to the lungs. Blood circulates back to the lungs and the process repeats.

Aerobic Respiration

Molecular oxygen, O_2, is essential for cellular respiration in all aerobic organisms. Oxygen is used as an electron acceptor in mitochondria to generate chemical energy in the form of adenosine triphosphate (ATP) during oxidative phosphorylation. The reaction for the aerobic respiration is essentially the reverse of photosynthesis, except that now there is a large release of chemical energy that is stored in ATP molecules (up to 38 ATP molecules are formed from one molecule of glucose). The simplified version of this reaction is:

$$C_6H_{12}O_6 + 6O_2 \rightarrow 6CO_2 + 6H_2O + 2880 \text{ kJ/mol}$$

Reactive Oxygen Species

Reactive oxygen species are dangerous by-products that sometimes result from the use of oxygen in organisms. Important examples include; oxygen free radicals such as the highly-dangerous superoxide O_2^-, and the less harmful hydrogen peroxide (H_2O_2). The body uses superoxide dismutase to reduce superoxide radicals to hydrogen peroxide. Glutathione peroxidase and similar enzymes then convert the H_2O_2 to water and dioxygen.

Parts of the immune system of higher organisms, however, create peroxide, superoxide, and singlet oxygen to destroy invading microbes. Recently, singlet oxygen has been found to be a source of biologically-produced ozone: This reaction proceeds through an unusual compound dihydrogen trioxide, also known as trioxidane, (HOOOH), which is an antibody-catalysed product of singlet oxygen and water. This compound, in turn, disproportionates to ozone and peroxide, providing two powerful antibacterials. The body's range of defence against all of these active oxidizing agents is hardly surprising, then, given their "deliberate" employment as antimicrobial agents in the immune response. Reactive oxygen species also play an important role in the hypersensitive response of plants against pathogen attack.

Electron Transfers

The electron is a subatomic particle carrying a negative electric charge. It has no known components or substructure. Therefore, the electron is generally believed to be an elementary particle. An electron has a mass that is approximately 1/1836 that of the proton. The intrinsic angular momentum (spin) of the electron is a half-integer value in units of ‘, which means that it is a fermion. The antiparticle of the electron is called the positron. The positron is identical to the electron except that it carries electrical and other charges of the opposite sign. When an electron collides with a positron, both particles may either scatter off each other or be totally annihilated, producing a pair (or more) of gamma ray photons. Electrons, which belong to the first generation of the lepton particle family, participate in gravitational, electromagnetic and weak interactions. Electrons, like all matter, have quantum mechanical properties of both particles and waves, so they can collide with other particles and be diffracted like light. However, this duality is best demonstrated in experiments with electrons, due to their tiny mass. Since an electron is a fermion, no two electrons can occupy the same quantum state, in accordance with the Pauli exclusion principle.

The concept of an indivisible amount of electric charge was theorized to explain the chemical properties of atoms, beginning in 1838 by British natural philosopher Richard Laming; the name *electron* was introduced for this charge in 1894 by Irish physicist George Johnstone Stoney. The electron was identified as a particle in 1897 by J. J. Thomson and his team of British physicists. In many physical phenomena, such as electricity, magnetism, and thermal conductivity, electrons play an essential role. An electron in motion relative to an observer generates a magnetic field, and will be deflected by external magnetic fields. When an electron is accelerated, it can absorb or radiate energy in the form of photons. Electrons, together with atomic nuclei made of protons and neutrons, make up atoms. However, electrons contribute less than 0.06% to an atom's total mass. The attractive Coulomb force between an electron and a proton causes electrons to be bound into atoms. The exchange or sharing of the electrons between two or more atoms is the main cause of chemical bonding.

According to theory, most electrons in the universe were created in the big bang, but they may also be created through beta decay of radioactive isotopes and in high-energy collisions, for instance when cosmic rays enter the atmosphere. Electrons may be destroyed through annihilation with positrons, and may be absorbed during nucleosynthesis in stars. Laboratory instruments are capable of containing and observing individual electrons as well as electron plasma, whereas dedicated telescopes can detect electron plasma in outer space. Electrons have many applications, including welding, cathode ray tubes, electron microscopes, radiation therapy, lasers and particle accelerators.

Atomic Theory

By 1914, experiments by physicists Ernest Rutherford, Henry Moseley, James Franck and Gustav Hertz had largely established the structure of an atom as a dense nucleus of positive charge surrounded by lower-mass electrons.

In 1913, Danish physicist Niels Bohr postulated that electrons resided in quantized energy states, with the energy determined by the angular momentum of the electron's orbits about the nucleus. The electrons could move between these states, or orbits, by the emission or absorption of photons at specific frequencies. By means of these quantized orbits, he accurately explained the spectral lines of the hydrogen atom. However, Bohr's model failed to account for the relative intensities of the spectral lines and it was unsuccessful in explaining the spectra of more complex atoms.

Chemical bonds between atoms were explained by Gilbert Newton Lewis, who in 1916 proposed that a covalent bond between two atoms is maintained by a pair of electrons shared between them. Later, in 1923, Walter Heitler and Fritz London gave the full explanation of the electron-pair formation and chemical bonding in terms of quantum mechanics. In 1919, the American chemist Irving Langmuir elaborated on the Lewis' static model of the atom and suggested that all electrons were distributed in successive "concentric (nearly) spherical shells, all of equal thickness". The shells were, in turn, divided by him in a number of cells each containing one pair of electrons. With this model Langmuir was able to qualitatively explain the chemical properties of all elements in the periodic table, which were known to largely repeat themselves according to the periodic law.

In 1924, Austrian physicist Wolfgang Pauli observed that the shell-like structure of the atom could be explained by a set of four parameters that defined every quantum energy state, as long as each state was inhabited by no more than a single electron. (This prohibition against more than one electron occupying the same quantum energy state became known as the Pauli exclusion principle.) The physical mechanism to explain the fourth parameter, which had two distinct possible values, was provided by the Dutch physicists Abraham Goudsmith and George Uhlenbeck when they suggested that an electron, in addition to the angular momentum of its orbit, could possess an intrinsic angular momentum. This property became known as spin, and explained the previously mysterious splitting of spectral lines observed with a high-resolution spectrograph; this phenomenon is known as fine structure splitting.

Quantum Mechanics

In his 1924 dissertation *Recherches sur la théorie des quanta* (Research on Quantum Theory), French physicist Louis de Broglie hypothesized that all matter possesses a De Broglie wave similar to light. That is, under the appropriate conditions, electrons and other matter would show properties of either particles or waves. The corpuscular properties of a particle are demonstrated when it is shown to have a localized position in space along its trajectory at any given moment. Wave-like nature is observed, for example, when a beam of light is passed through parallel slits and creates interference patterns. In 1927, the interference effect was demonstrated with a beam of electrons by English physicist George Paget Thomson with a thin metal film and by American physicists Clinton Davisson and Lester Germer using a crystal of nickel.

The success of de Broglie's prediction led to the publication, by Erwin Schrodinger in 1926, of the Schrodinger equation that successfully describes how electron waves propagated. Rather than yielding a solution that determines the location of an electron over time, this wave equation can be used to predict the probability of finding an electron near a position. This approach was later called quantum mechanics, which provided an extremely close derivation to the energy states of an electron in a hydrogen atom. Once spin and the interaction between multiple electrons were considered, quantum mechanics allowed the configuration of electrons in atoms with higher atomic numbers than hydrogen to be successfully predicted.

In 1928, building on Wolfgang Pauli's work, Paul Dirac produced a model of the electron-the Dirac equation, consistent with relativity theory, by applying relativistic and symmetry considerations to the hamiltonian formulation of the quantum mechanics of the electro-magnetic field. In order to resolve some problems within his relativistic equation, in 1930 Dirac developed a model of the vacuum as an infinite sea of particles having negative energy, which was dubbed the Dirac sea. This led him to predict the existence of a positron, the antimatter counterpart of the electron. This particle was discovered in 1932 by Carl D. Anderson, who proposed calling standard electrons *negatrons*, and using *electron* as a generic term to describe both the positively and negatively charged variants. This usage of the term 'negatron' is still occasionally encountered today, and it may be shortened to 'negaton'.

In 1947 Willis Lamb, working in collaboration with graduate student Robert Rutherford, found that certain quantum states of hydrogen atom, which should have the same energy, were shifted in relation to each other, the difference being the Lamb shift. About the same time, Polykarp Kusch, working with Henry M. Foley, discovered the magnetic moment of the electron is slightly larger than predicted by Dirac's theory. This small difference was later called anomalous magnetic dipole moment of the electron. To resolve these issues, a refined theory called quantum electrodynamics was developed by Sin-Itiro Tomonaga, Julian Schwinger and Richard P. Feynman in the late 1940s.

Particle Accelerators

With the development of the particle accelerator during the first half of the twentieth century, physicists began to delve deeper into the properties of subatomic particles. The first successful attempt to accelerate electrons using Electromagnetic induction was made in 1942 by Donald Kerst. His initial betatron reached energies of 2.3 MeV, while subsequent betatrons achieved 300 MeV. In 1947, synchrotron radiation was discovered with a 70 MeV electron synchrotron at General Electric. This radiation was caused by the acceleration of electrons, moving near the speed of light, through a magnetic field.

With a beam energy of 1.5 GeV, the first high-energy particle collider was ADONE, which began operations in 1968. This device accelerated electrons and positrons in opposite directions, effectively doubling the energy of their collision when compared to striking a static target with an electron. The Large Electron-Positron Collider (LEP) at CERN, which was operational from 1989 to 2000, achieved collision energies of 209 GeV and made important measurements for the Standard Model of particle physics.

Characteristics

Classification

In the Standard Model of particle physics, electrons belong to the group of subatomic particles called leptons, which are believed to be fundamental or elementary particles. Electrons have the lowest mass of any charged lepton (or electrically charged particle of any type) and belong to the first-generation of fundamental particles. The second and third generation contain charged leptons, the muon and the tau, which are identical to the electron in charge, spin and interactions, but are more massive. Leptons differ from the other basic constituent of matter, the quarks, by their lack of strong interaction. All members of the lepton group are fermions, because they all have half-odd integer spin; the electron has spin $^1D_2$.

Fundamental Properties

The invariant mass of an electron is approximately 9.109×10^{-31} kilogram, or 5.489×10^{-4} atomic mass unit. On the basis of Einstein's principle of mass–energy equivalence, this mass corresponds to a rest energy of 0.511 MeV. The ratio between the mass of a proton and that of an electron is about 1836. Astronomical measurements show that the proton-to-electron mass ratio has held the same value for at least half the age of the universe, as is predicted by the Standard Model.

Electrons have an electric charge of "1.602×10^{-19} coulomb, which is used as a standard unit of charge for subatomic particles. Within the limits of experimental accuracy, the electron charge is identical to the charge of a proton, but with the opposite sign. As the symbol *e* is used for the elementary charge, the electron is commonly symbolized by e^-, where the minus sign indicates the negative charge. The positron is symbolized by e^+ because it has the same properties as the electron but with a positive rather than negative charge.

The electron has an intrinsic angular momentum or spin of $^1D_2$. This property is usually stated by referring to the electron as a spin-$^1D_2$ particle. For such particles the spin magnitude is "$^3D_2$'. while the result of the measurement of a projection of the spin on any axis can only be $\pm D_2$. In addition to spin, the electron has an intrinsic magnetic moment along its spin axis.

It is approximately equal to one Bohr magneton, which is a physical constant equal to $9.274\ 009\ 15(23) \times 10^{-24}$ joules per tesla. The orientation of the spin with respect to the momentum of the electron defines the property of elementary particles known as helicity.

The electron has no known substructure. Hence, it is defined or assumed to be a point particle with a point charge and no spatial extent. Observation of a single electron in a Penning trap shows the upper limit of the particle's radius is 10^{-22} meters. There *is* a physical constant called the "classical electron radius", with the much larger value of 2.8179×10^{-15} m. However, the terminology comes from a simplistic calculation that ignores the effects of quantum mechanics; in reality, the so-called classical electron radius has little to do with the true fundamental structure of the electron.

There are elementary particles that spontaneously decay into less massive particles. An example is the muon, which decays into an electron, a neutrino and an antineutrino, with a mean lifetime of 2.2×10^{-6} seconds. However, the electron is thought to be stable on theoretical grounds: the electron is the least massive particle with non-zero electric charge, so its decay would violate charge conservation. The experimental lower bound for the electron's mean lifetime is 4.6×10^{26} years, at a 90% confidence level.

Quantum Properties

As with all particles, electrons can act as waves. This is called the wave–particle duality and can be demonstrated using the double-slit experiment. The wave-like nature of the electron allows it to pass through two parallel slits simultaneously, rather than just one slit as would be the case for a classical particle. In quantum mechanics, the wave-like property of one particle can be described mathematically as a complex-valued function, the wave function, commonly denoted by the Greek letter psi (ø). When the absolute value of this function is squared, it gives the probability that a particle will be observed near a location—a probability density.

Electrons are identical particles because they cannot be distinguished from each other by their intrinsic physical properties. In quantum mechanics, this means that a pair of interacting electrons must be able to swap positions without an observable change to the state of the system. The wave function of fermions, including electrons, is antisymmetric, meaning that it changes sign when two electrons are swapped; that is, $ø(r_1, r_2)$ = "$ø(r_2, r_1)$, where the variables r_1 and r_2 correspond to the first and second electrons, respectively. Since the absolute value is not changed by a sign swap, this corresponds to equal probabilities. Bosons, such as the photon, have symmetric wave functions instead.

In the case of antisymmetry, solutions of the wave equation for interacting electrons result in a zero probability that each pair will occupy the same location or state. This is responsible for the Pauli exclusion principle, which precludes any two electrons from occupying the same quantum state. This principle explains many of the properties of electrons. For example, it causes groups of bound electrons to occupy different orbitals in an atom, rather than all overlapping each other in the same orbit.

Virtual Particles

Physicists believe that empty space may be continually creating pairs of virtual particles, such as a positron and electron, which rapidly annihilate each other shortly thereafter. The combination of the energy variation needed to create these particles, and the time during which they exist, fall under the threshold of detectability expressed by the Heisenberg uncertainty relation, $\Delta E \cdot \Delta t$ e" '. In effect, the energy needed to create these virtual particles, ΔE, can be "borrowed" from the vacuum for a period of time, Δt, so that their product is no more than the reduced Planck constant, ' H- 6.6×10^{-16} eV·s. Thus, for a virtual electron, Δt is at most 1.3×10^{-21} s.

While an electron–positron virtual pair is in existence, the coulomb force from the ambient electric field surrounding an electron causes a created positron to be attracted to the original electron, while a created electron experiences a repulsion. This causes what is called vacuum polarization. In effect, the vacuum behaves like a medium having a dielectric permittivity more than unity. Thus the effective charge of an electron is actually smaller than its true value, and the charge decreases with increasing distance from the electron. This polarization was confirmed experimentally in 1997 using the Japanese TRISTAN particle accelerator. Virtual particles cause a comparable shielding effect for the mass of the electron.

The interaction with virtual particles also explains the small (about 0.1%) deviation of the intrinsic magnetic moment of the electron from the Bohr magneton (the anomalous magnetic moment). The extraordinarily precise agreement of this predicted difference with the experimentally determined value is viewed as one of the great achievements of quantum electrodynamics.

In classical physics, the angular momentum and magnetic moment of an object depend upon its physical dimensions. Hence, the concept of a dimensionless electron possessing these properties might seem inconsistent. The apparent paradox can be explained by the formation of virtual photons in the electric field generated by the electron. These photons cause the electron to shift about in a jittery fashion (known as zitterbewegung), which results in a net circular motion with precession. This motion produces both the spin and the magnetic moment of the electron. In atoms, this creation of virtual photons explains the Lamb shift observed in spectral lines.

Interaction

An electron generates an electric field that exerts an attractive force on a particle with a positive charge, such as the proton, and a repulsive force on a particle with a negative charge. The strength of this force is determined by Coulomb's inverse square law. When an electron is in motion, it generates a magnetic field. The Ampère-Maxwell law relates the magnetic field to the mass motion of electrons (the current) with respect to an observer. It is this property of induction which supplies the magnetic field that drives an electric motor. The electromagnetic field of an arbitrary moving charged particle is expressed by the Liénard–Wiechert potentials, which are valid even when the particle's speed is close to that of light (relativistic).

When an electron is moving through a magnetic field, it is subject to the Lorentz force that exerts an influence in a direction perpendicular to the plane defined by the magnetic field and the electron velocity. This centripetal force causes the electron to follow a helical trajectory through the field at a radius called the gyroradius.

The acceleration from this curving motion induces the electron to radiate energy in the form of synchrotron radiation. The energy emission in turn causes a recoil of the electron, known as the Abraham-Lorentz-

Dirac force, which creates a friction that slows the electron. This force is caused by a back-reaction of the electron's own field upon itself.

In quantum electrodynamics the electromagnetic interaction between particles is mediated by photons. An isolated electron that is not undergoing acceleration is unable to emit or absorb a real photon; doing so would violate conservation of energy and momentum. Instead, virtual photons can transfer momentum between two charged particles. It is this exchange of virtual photons that, for example, generates the Coulomb force. Energy emission can occur when a moving electron is deflected by a charged particle, such as a proton. The acceleration of the electron results in the emission of Bremsstrahlung radiation.

An inelastic collision between a photon (light) and a solitary (free) electron is called Compton scattering. This collision results in a transfer of momentum and energy between the particles, which modifies the wavelength of the photon by an amount called the Compton shift. The maximum magnitude of this wavelength shift is $h/m_e c$, which is known as the Compton wavelength. For an electron, it has a value of 2.43×10^{-12} m. When the wavelength of the light is long (for instance, the wavelength of the visible light is 0.4–0.7 μm) the wavelength shift becomes negligible. Such interaction between the light and free electrons is called Thomson scattering or Linear Thomson scattering.

The relative strength of the electromagnetic interaction between two charged particles, such as an electron and a proton, is given by the fine-structure constant. This value is a dimensionless quantity formed by the ratio of two energies: the electrostatic energy of attraction (or repulsion) at a separation of one Compton wavelength, and the rest energy of the charge. It is given by α H– 7.297353×10^{-3}, which is approximately equal to $^{1}D_{137}$.

When electrons and positrons collide, they annihilate each other, giving rise to two or more gamma ray photons. If the electron and positron have negligible momentum, a positronium atom can form before annihilation results in two or three gamma ray photons totalling 1.022 MeV. On the other hand, high-energy photons may transform into an electron and a positron by a process called pair production, but only in the presence of a nearby charged particle, such as a nucleus.

In the theory of electroweak interaction, the left-handed component of electron's wavefunction forms a weak isospin doublet with the electron neutrino. This means that during weak interactions, electron neutrinos behave like electrons. Either member of this doublet can undergo a charged current interaction by emitting or absorbing a W and be converted into the other member. Charge is conserved during this reaction because the

W boson also carries a charge, canceling out any net change during the transmutation. Charged current interactions are responsible for the phenomenon of beta decay in a radioactive atom. Both the electron and electron neutrino can undergo a neutral current interaction via a Z^0 exchange, and this is responsible for neutrino-electron elastic scattering.

Atoms and Molecules

Probability densities for the first few hydrogen atom orbitals, seen in cross-section. The energy level of a bound electron determines the orbital it occupies, and the colour reflects the probability to find the electron at a given position.

An electron can be *bound* to the nucleus of an atom by the attractive Coulomb force. A system of several electrons bound to a nucleus is called an atom. If the number of electrons is different from the nucleus' electrical charge, such an atom is called an ion. The wave-like behaviour of a bound electron is described by a function called an atomic orbital. Each orbital has its own set of quantum numbers such as energy, angular momentum and projection of angular momentum, and only a discrete set of these orbitals exist around the nucleus. According to the Pauli exclusion principal each orbital can be occupied by up to two electrons, which must differ in their spin quantum number.

Electrons can transfer between different orbitals by the emission or absorption of photons with an energy that matches the difference in potential. Other methods of orbital transfer include collisions with particles, such as electrons, and the Auger effect. In order to escape the atom, the energy of the electron must be increased above its binding energy to the atom. This occurs, for example, with the photoelectric effect, where an incident photon exceeding the atom's ionization energy is absorbed by the electron.

The orbital angular momentum of electrons is quantized. Because the electron is charged, it produces an orbital magnetic moment that is proportional to the angular momentum. The net magnetic moment of an atom is equal to the vector sum of orbital and spin magnetic moments of all electrons and the nucleus. The nuclear magnetic moment is, however, negligible in comparison to the effect from the electrons. The magnetic moments of the electrons that occupy the same orbital (so called, paired electrons) cancel each other out.

The chemical bond between atoms occurs as a result of electromagnetic interactions, as described by the laws of quantum mechanics. The strongest bonds are formed by the sharing or transfer of electrons between atoms, allowing the formation of molecules. Within a molecule, electrons move under the influence of several nuclei, and occupy molecular orbitals; much as they can occupy atomic orbitals in isolated atoms. A fundamental factor in these molecular structures is the existence of electron pairs. These are electrons with opposed spins, allowing them to occupy the same molecular orbital without violating the Pauli exclusion principle (much like in atoms). Different molecular orbitals have different spatial distribution of the electron density. For instance, in bonded pairs (i.e. in the pairs that actually bind atoms together) electrons can be found with the maximal probability in a relatively small volume between the nuclei. On the contrary, in non-bonded pairs electrons are distributed in a large volume around nuclei.

Conductivity

If a body has more or fewer electrons than are required to balance the positive charge of the nuclei, then that object has a net electric charge. When there is an excess of electrons, the object is said to be negatively charged. When there are fewer electrons than the number of protons in nuclei, the object is said to be positively charged. When the number of electrons and the number of protons are equal, their charges cancel each other and the object is said to be electrically neutral. A macroscopic body can develop an electric charge through rubbing, by the triboelectric effect.

Independent electrons moving in vacuum are termed *free* electrons. Electrons in metals also behave as if they were free. In reality the particles that are commonly termed electrons in metals and other solids are quasi-electrons—quasi-particles, which have the same electrical charge, spin and magnetic moment as real electrons but may have a different mass. When free electrons—both in vacuum and metals—move, they produce a net flow of charge called an electric current, which generates a magnetic field. Likewise a current can be created by a changing magnetic field. These interactions are described mathematically by Maxwell's equations.

At a given temperature, each material has an electrical conductivity that determines the value of electric current when an electric potential is applied. Examples of good conductors include metals such as copper and gold, whereas glass and Teflon are poor conductors. In any dielectric material, the electrons remain bound to their respective atoms and the material behaves as an insulator. Most semiconductors have a variable level of conductivity that lies between the extremes of conduction and insulation. On the other hand, metals have an electronic band structure containing partially filled electronic bands. The presence of such bands allows electrons in metals to behave as if they were free or delocalized electrons. These electrons are not associated with specific atoms, so when an electric field is applied, they are free to move like a gas (called Fermi gas) through the material much like free electrons.

Because of collisions between electrons and atoms, the drift velocity of electrons in a conductor is on the order of millimeters per second. However, the speed at which a change of current at one point in the material causes changes in currents in other parts of the material, the velocity of propagation, is typically about 75% of light speed. This occurs because electrical signals propagate as a wave, with the velocity dependent on the dielectric constant of the material.

Metals make relatively good conductors of heat, primarily because the delocalized electrons are free to transport thermal energy between atoms. However, unlike electrical conductivity, the thermal conductivity of a metal is nearly independent of temperature. This is expressed mathematically by the Wiedemann-Franz law, which states that the ratio of thermal conductivity to the electrical conductivity is proportional to the temperature. The thermal disorder in the metallic lattice increases the electrical resistivity of the material, producing a temperature dependence for electrical current.

When cooled below a point called the critical temperature, materials can undergo a phase transition in which they lose all resistivity to electrical current, in a process known as superconductivity. In BCS theory, this behaviour is modeled by pairs of electrons entering a quantum state known as a Bose–Einstein condensate. These Cooper pairs have their motion coupled to nearby matter via lattice vibrations called phonons, thereby avoiding the collisions with atoms that normally create electrical resistance. (Cooper pairs have a radius of roughly 100 nm, so they can overlap each other.) However, the mechanism by which higher temperature superconductors operate remains uncertain.

Electrons inside conducting solids, which are quasi-particles themselves, when tightly confined at temperatures close to absolute zero, behave as though they had split into two other quasiparticles: spinons and holons. The former carries spin and magnetic moment, while the latter electrical charge.

Motion and Energy

According to Einstein's theory of special relativity, as an electron's speed approaches the speed of light, from an observer's point of view its relativistic mass increases, thereby making it more and more difficult to accelerate it from within the observer's frame of reference. The speed of an electron can approach, but never reach, the speed of light in a vacuum, *c*. However, when relativistic electrons—that is, electrons moving at a speed close to *c*—are injected into a dielectric medium such as water, where the local speed of light is significantly less than *c*, the electrons temporarily travel faster than light in the medium. As they interact with the medium, they generate a faint light called Cherenkov radiation.

The effects of special relativity are based on a quantity known as the Lorentz factor, defined as $\gamma=1/\sqrt{1-v^2/c^2}$ where v is the speed of the particle. The kinetic energy K_e of an electron moving with velocity v is:

$$K_e = (\gamma - 1)m_e c^2.$$

where m_e is the mass of electron. For example, the Stanford linear accelerator can accelerate an electron to roughly 51 GeV. This gives a value of nearly 100,000 for γ, since the mass of an electron is 0.51 MeV/c^2. The relativistic momentum of this electron is 100,000 times the momentum that classical mechanics would predict for an electron at the same speed. Since an electron behaves as a wave, at a given velocity it has a characteristic de Broglie wavelength. This is given by $\lambda_e = h/p$ where h is the Planck constant and p is the momentum. For the 51 GeV electron above, the wavelength is about 2.4×10^{-17} m, small enough to explore structures well below the size of an atomic nucleus.

Formation

The Big Bang theory is the most widely accepted scientific theory to explain the early stages in the evolution of the Universe. For the first millisecond of the Big Bang, the temperatures were over 10 billion kelvins and photons had mean energies over a million electronvolts. These photons were sufficiently energetic that they could react with each other to form pairs of electrons and positrons,

$$\gamma + \gamma \rightleftharpoons e^+ + e^-.$$

where γ is a photon, e^+ is a positron and e^- is an electron. Likewise, positron-electron pairs annihilated each other and emitted energetic photons. An equilibrium between electrons, positrons and photons was maintained during this phase of the evolution of the Universe. After 15 seconds had passed, however, the temperature of the universe dropped below the threshold where electron-positron formation could occur. Most of the surviving electrons and positrons annihilated each other, releasing gamma radiation that briefly reheated the universe. For reasons that remain uncertain, during the process of leptogenesis there was an excess in the number of electrons over positrons. Hence, about one electron in every billion survived the annihilation process. This excess matched the excess of protons over anti-protons, in a condition known as baryon asymmetry, resulting in a net charge of zero for the universe.

The surviving protons and neutrons began to participate in reactions with each other—in the process known as nucleosynthesis, forming isotopes of hydrogen and helium, with trace amounts of lithium. This process peaked after about five minutes. Any leftover neutrons underwent negative beta decay with a half-life of about a thousand seconds, releasing a proton and electron in the process,

$$n \Rightarrow p + e^- + \bar{\nu}_e:$$

where n is a neutron, p is a proton and ν_e is an electron antineutrino. For about the next 300,000–400,000 years, the excess electrons remained too energetic to bind with atomic nuclei. What followed is a period known as recombination, when neutral atoms were formed and the expanding universe became transparent to radiation. Roughly one million years after the big bang, the first generation of stars began to form. Within a star, stellar nucleosynthesis results in the production of positrons from the fusion of atomic nuclei. These antimatter particles immediately annihilate with electrons, releasing gamma rays. The net result is a steady reduction in the number of electrons, and a matching increase in the number of neutrons. However, the process of stellar evolution can result in the synthesis of radioactive isotopes. Selected isotopes can subsequently undergo negative beta decay, emitting an electron and antineutrino from the nucleus. An example is the cobalt-60 (^{60}Co) isotope, which decays to form nickel-60 (^{60}Ni).

At the end of its lifetime, a star with more than about 20 solar masses can undergo gravitational collapse to form a black hole. According to classical physics, these massive stellar objects exert a gravitational attraction that is strong enough to prevent anything, even electromagnetic radiation, from escaping past the Schwarzschild radius. However, it is believed that quantum mechanical effects may allow Hawking radiation to be emitted at this distance. Electrons (and positrons) are thought to be created at the event horizon of these stellar remnants. When pairs of virtual particles (such as an electron and positron) are created in the vicinity of the event horizon, the random spatial distribution of these particles may permit one of them to appear on the exterior; this process is called quantum tunnelling. The gravitational potential of the black hole can then supply the energy that transforms this virtual particle into a real particle, allowing it to radiate away into space. In exchange, the other member of the pair is given negative energy, which results in a net loss of mass-energy by the black hole. The rate of Hawking radiation increases with decreasing mass, eventually causing the black hole to evaporate away until, finally, it explodes.

Cosmic rays are particles travelling through space with high energies. Energy events as high as 3.0×10^{20} eV have been recorded. When these particles collide with nucleons in the Earth's atmosphere, a shower of particles is generated, including pions. More than half of the cosmic radiation observed from the Earth's surface consists of muons. The particle called a muon is a lepton which is produced in the upper atmosphere by the decay of a pion. A muon, in turn, can decay to form an electron or positron. Thus, for the negatively charged pion π^-,

$$\pi^- \rightarrow \mu^- + \bar{\nu}_\mu:$$

$$\mu^- \rightarrow e^- + \bar{\nu}_e + \nu_\mu:$$

where μ^- is a muon and υ_μ is a muon neutrino.

Observation

Remote observation of electrons requires detection of their radiated energy. For example, in high-energy environments such as the corona of a star, free electrons form a plasma that radiates energy due to Bremsstrahlung. Electron gas can undergo plasma oscillation, which is waves caused by synchronized variations in electron density, and these produce energy emissions that can be detected by using radio telescopes. The frequency of a photon is proportional to its energy. As a bound electron transitions between different energy levels of an atom, it will absorb or emit photons at characteristic frequencies. For instance, when atoms are irradiated by a source with a broad spectrum, distinct absorption lines will appear in the spectrum of transmitted radiation. Each element or molecule displays a characteristic set of spectral lines, such as the hydrogen spectral series. Spectroscopic measurements of the strength and width of these lines allow the composition and physical properties of a substance to be determined.

In laboratory conditions, the interactions of individual electrons can be observed by means of particle detectors, which allow measurement of specific properties such as energy, spin and charge. The development of the Paul trap and Penning trap allows charged particles to be contained within a small region for long durations. This enables precise measurements of the particle properties. For example, in one instance a Penning trap was used to contain a single electron for a period of 10 months. The magnetic moment of the electron was measured to a precision of eleven digits, which, in 1980, was a greater accuracy than for any other physical constant.

The first video images of an electron's energy distribution were captured by a team at Lund University in Sweden, February 2008. The scientists used extremely short flashes of light, called attosecond pulses, which allowed an electron's motion to be observed for the first time. The distribution of the electrons in solid materials can be visualized by angle resolved photoemission spectroscopy (ARPES). This technique employs the photoelectric effect to measure the reciprocal space—a mathematical representation of periodic structures that is used to infer the original structure. ARPES can be used to determine the direction, speed and scattering of electrons within the material.

Plasma Applications

Particle Beams

Electron beams are used in welding, which allows energy densities up to 10^7 W•cm^{-2} across a narrow focus diameter of 0.1–1.3 mm and usually does not require a filler material. This welding technique must be performed in a vacuum, so that the electron beam does not interact with the gas prior to reaching the target, and it can be used to join conductive materials that would otherwise be considered unsuitable for welding. Electron beam lithography (EBL) is a method of etching semiconductors at resolutions smaller than a micron. This technique is limited by high costs, slow performance, the need to operate the beam in the vacuum and the tendency of the electrons to scatter in solids. The last problem limits the resolution to about 10 nm. For this reason, EBL is primarily used for the production of small numbers of specialized integrated circuits. Electron beam processing is used to irradiate materials in order to change their physical properties or sterilize medical and food products. In radiation therapy, electron beams are generated by linear accelerators for treatment of superficial tumours. Because an electron beam only penetrates to a limited depth before being absorbed, typically up to 5 cm for electron energies in the range 5–20 MeV, electron therapy is useful for treating skin lesions such as basal cell carcinomas. An electron beam can be used to supplement the treatment of areas that have been irradiated by X-rays.

Particle accelerators use electric fields to propel electrons and their antiparticles to high energies. As these particles pass through magnetic fields, they emit synchrotron radiation. The intensity of this radiation is spin dependent, which causes polarization of the electron beam—a process known as the Sokolov–Ternov effect. The polarized electron beams can be useful for various experiments. Synchrotron radiation can also be used for cooling the electron beams, which reduces the momentum spread of the particles. Once the particles have accelerated to the required energies, separate electron and positron beams are brought into collision. The resulting energy emissions are observed with particle detectors and are studied in particle physics.

Imaging

Low-energy electron diffraction (LEED) is a method of bombarding a crystalline material with a collimated beam of electrons, then observing the resulting diffraction patterns to determine the structure of the material. The required energy of the electrons is typically in the range 20–200 eV. The reflection high energy electron diffraction (RHEED) technique uses the reflection of a beam of electrons fired at various low angles to characterize the surface of crystalline materials. The beam energy is typically in the range 8–20 keV and the angle of incidence is 1–4°.

The electron microscope directs a focused beam of electrons at a specimen. As the beam interacts with the material, some electrons change their properties, such as movement direction, angle, relative phase and energy. By recording these changes in the electron beam, microscopists can produce atomically resolved image of the material. In blue light, conventional optical microscopes have a diffraction-limited resolution of about 200 nm. By comparison, electron microscopes are limited by the de Broglie wavelength of the electron. This wavelength, for example, is equal to 0.0037 nm for electrons accelerated across a 100,000-volt potential. The Transmission Electron Aberration-corrected Microscope is capable of sub-0.05 nm resolution, which is more than enough to resolve individual atoms. This capability makes the electron microscope a useful laboratory instrument for high resolution imaging. However, electron microscopes are expensive instruments that are costly to maintain. There are two main types of electron microscopes: transmission and scanning. Transmission electron microscopes function in a manner similar to overhead projector, with a beam of electrons passing through a slice of material then being projected by lenses on a photographic slide or a charge-coupled device. In scanning electron microscopes, the image is produced by rastering a finely focused electron beam, as in a TV set, across the studied sample. The magnifications range from 100× to 1,000,000× or higher for both microscope types. The scanning tunnelling microscope uses quantum tunnelling of electrons from a sharp metal tip into the studied material and can produce atomically resolved images of its surface.

Other

In the free electron laser (FEL), a relativistic electron beam is passed through a pair of undulators containing arrays of dipole magnets, whose fields are oriented in alternating directions. The electrons emit synchrotron radiation, which, in turn, coherently interacts with the same electrons. This leads to the strong amplification of the radiation field at the resonance frequency. FEL can emit a coherent high-brilliance electromagnetic radiation with a wide range of frequencies, from microwaves to soft X-rays. These devices can be used in the future for manufacturing, communication and various medical applications, such as soft tissue surgery.

Electrons are at the heart of cathode ray tubes, which are used extensively as display devices in laboratory instruments, computer monitors and television sets. In a photomultiplier tube, every photon striking the photocathode initiates an avalanche of electrons that produces a detectable current pulse. Vacuum tubes use the flow of electrons to manipulate electrical signals, and they played a critical role in the development of electronics technology. However, they have been largely supplanted by solid-state devices such as the transistor.

Electron Transfer

Electron transfer (ET) is the process by which an electron moves from one atom or molecule to another atom or molecule. ET is a mechanistic description of the thermodynamic concept of redox, wherein the oxidation states of both reaction partners change. Numerous processes in biology involve ET reactions including oxygen binding, photosynthesis, respiration, and detoxification routes. Additionally, the process of energy transfer can be formalized as a two electron exchange (two concurrent ET events in opposite directions). ET reactions commonly involve transition metal complexes, but there are now many examples of ET in organic chemistry.

Classes of Electron Transfer

There are several classes of electron transfer, defined by the state of the two redox centres and their connectivity

Inner-sphere Electron Transfer

In inner-sphere ET, the two redox centres are covalently linked during the ET. This bridge can be permanent, in which case the electron transfer event is termed intramolecular electron transfer. More commonly, however, the covalent linkage is transitory, forming just prior to the ET and then disconnecting following the ET event. In such cases, the electron transfer is termed intermolecular electron transfer. A famous example of an inner sphere ET process that proceeds via a transitory bridged intermediate is the reduction of $[CoCl(NH_3)_5]^{2+}$ by $[Cr(H_2O)_6]^{2+}$. In this case the chloride ligand is the bridging ligand that covalently connects the redox partners.

Inner sphere or bonded electron transfer proceeds via a covalent linkage between the two redox partners, the oxidant and the reductant. In Inner Sphere (IS) electron transfer (ET), a ligand bridges the two metal redox centres during the electron transfer event. Inner sphere reactions are inhibited by large ligands, which prevent the formation of the crucial bridged intermediate. Thus, IS ET is rare in biological systems, where redox sites are often shielded by bulky proteins. Inner sphere ET is usually used to describe reactions involving transition metal complexes and most of this article is written from this perspective. However, redox centres can consist of organic groups rather than metal centres.

The bridging ligand could be virtually any entity that can convey electrons. Typically, such a ligand has more than one lone electron pair, such that it can serve as an electron donor to both the reductant and the oxidant. Common bridging ligands include the halides and the pseudohalides such as hydroxide and thiocyanate. More complex bridging ligands are also well known including oxalate, malonate, and pyrazine. Prior to ET, the bridged complex must form, and such processes are often highly reversible.

Electron transfer occurs through the bridge once it is established. In some cases, the a stable bridged structure may exist in the ground state; in other cases, the bridged structure may be a transiently-formed intermediate, or else as a transition state during the reaction. The alternative to inner sphere electron transfer is outer sphere electron transfer. In any transition metal redox process, the mechanism can be assumed to be outer sphere unless the conditions of the inner sphere are met. Inner sphere electron transfer is generally enthalpically more favourable than outer sphere electron transfer due to a larger degree of interaction between the metal centres involved, however, inner sphere electron transfer is usually entropically less favourable since the two sites involved must become more ordered (come together via a bridge) than in outer sphere electron transfer.

Outer-sphere Electron Transfer

Outer sphere or nonbonded electron transfer refers to an electron transfer (ET) event that occurs between chemical moieties that remain separate species before, during, and after the ET event. This is opposed to inner sphere electron transfer in which the two sites undergoing ET are connected by a chemical bridge. Because the ET in outer sphere electron transfer occurs between two non-connected species, the electron is forced to move through space from one redox centre to the other. Outer sphere electron transfer can occur between different chemicals or between two chemicals that differ only in their oxidation state. The latter is termed "self-exchange".

Outer sphere electron transfer is generally enthalpically less favourable than inner sphere electron transfer because the interaction through space between the redox centres in outersphere electron transfer is weaker than the interaction through the chemical bridge present in the inner sphere mechanism. By the same token, outer sphere electron transfer is usually entropically more favourable than inner sphere electron transfer as the two sites involved do not have to go through the ordering processes associated with the formation of a bridge. The first generally accepted theory of electron transfer was developed to explain outer-sphere electron transfer. It was developed by Rudolph A. Marcus in the 1950s. Marcus theory is still the most widely accepted theory of outer-sphere electron transfer. The dependence of the electron transfer rate on the thermodynamic driving force (difference in the redox potentials of the electron-exchanging sites), including the so-called "Marcus inverted region", has also been interpreted by the catastrophe theory (Xu, F. Application of catastrophe theory to the "G`⁻ to -"G relationship in electron transfer reactions. Outer sphere electron transfer is the most commonly known type of electron transfer.

It is by far the more common form of electron transfer biochemistry, where the sterics of large proteins inhibit the formation of a chemical bridge between their redox centres. In biochemistry, there are two main types of outer sphere ET: ET between two biological molecules or fixed distance ET, in which the electron transfers within a *single* biomolecule (e.g. intraprotein).

In outer-sphere ET reactions, the participating redox centres are not linked via any bridge during the ET event. Instead, the electron "hops" through space from the reducing centre to the acceptor. Outer-sphere ET is by definition intermolecular. Outer sphere electron transfer can occur between different chemical species or between identical chemical species that differ only in their oxidation state. The later process is termed self-exchange. As an example, self-exchange describes the degenerate reaction between Permanganate and its one-electron reduced relative manganate:

$$[MnO_4]^- + [Mn^*O_4]^{2-} \rightarrow [MnO_4]^{2-} + [Mn^*O_4]^-$$

In general, if electron transfer is faster than ligand substitution, the reaction will follow the outer-sphere electron transfer.

Often occurs when one/both reactants are inert or if there is no suitable bridging ligand.

A key concept of Marcus theory is that the rates of such self-exchange reactions are mathematically related to the rates of "cross reactions". Cross reactions entail partners that differ by more than their oxidation states. One example (of many thousands) is the reduction of permanganate by iodide to form iodine and, again, manganate.

Five steps of an Outer Sphere Reaction

1. reactants differ together in solution => precursor complex (requires work $=w_r$)
2. changing bond lengths, reorganize solvent => activated complex
3. Electron transfer
4. Relaxation of bond lengths, solvent molecules => successor complex
5. Diffusion of products (requires work$=w_p$).

Heterogeneous Electron Transfer

In heterogeneous electron transfer, an electron moves between a chemical species and a solid-state electrode. Theories addressing heterogeneous electron transfer have applications in electrochemistry and the design of solar cells.

8

The Catalysts of Life

Enzymes: The Catalysts of Life

Each cell in the body is a beehive of activity called metabolism. The biochemical activity is the process of chemical and physical change which goes on continually in the living organism. These characteristics of life include the build-up of new tissue, replacement of old tissue, conversion of food to energy, disposal of waste materials, and reproduction.

This building up and tearing down takes place in the face of an apparent paradox. The greatest majority of these biochemical reactions do not take place spontaneously. The phenomenon of catalysis makes possible biochemical reactions necessary for all life processes. Catalysis is defined as the acceleration of a chemical reaction by some substance which itself undergoes no permanent chemical change.

The catalysts of biochemical reactions are enzymes and are responsible for bringing about almost all of the chemical reactions in living organisms. Without enzymes, these reactions take place at a rate far too slow for the pace of metabolism.

The oxidation of a fatty acid to carbon dioxide and water is not a gentle process in a test tube – extremes of pH, high temperatures and corrosive chemicals are required. Yet in the body, such a reaction takes place smoothly and rapidly within a narrow range of pH and temperature. In the laboratory, the average protein must be boiled for about 24 hours in a 20% HCl solution to achieve a complete breakdown. In the body, the breakdown takes place in four hours or less under conditions of mild physiological temperature and pH.

It is through attempts at understanding more about enzyme catalysts – what they are, what they do, and how they do it– that many advances in medicine and the life sciences have been brought about.

All known enzymes are proteins. They are high molecular weight compounds made up principally of chains of amino acids linked together by peptide bonds.

The primary benefit of enzymes for us is in digestion...

Digestive Enzymes

Specific enzymes work on specific foods. You need the right type of enzyme for the foods you want it to break down. Digestive enzymes are enzymes that break down food into usable material. The major different types of digestive enzymes are:

- amylase – breaks down carbohydrates, starches, and sugars which are prevalent in potatoes, fruits, vegetables, and many snack foods
- protease – breaks down proteins found in meats, nuts, eggs, and cheese
- lipase – breaks down fats found in most dairy products, nuts, oils, and meat
- cellulase – breaks down cellulose, plant fiber; not found in humans

Enzyme is a Biological Catalyst

Catalyst is a molecule that increases the rate of a chemical reaction without itself being permanently changed after the reaction. Enzymes are proteins that catalyse (*i.e.*, increase or decrease the rates of) chemical reactions. In enzymatic reactions, the molecules at the beginning of the process are called substrates, and they are converted into different molecules, called the products. Almost all processes in a biological cell need enzymes to occur at significant rates. Since enzymes are selective for their substrates and speed up only a few reactions from among many possibilities, the set of enzymes made in a cell determines which metabolic pathways occur in that cell.

Like all catalysts, enzymes work by lowering the activation energy ($E_a^{\ddagger}$) for a reaction, thus dramatically increasing the rate of the reaction. As a result, products are formed faster and reactions reach their equilibrium state more rapidly. Most enzyme reaction rates are millions of times faster than those of comparable un-catalysed reactions. As with all catalysts, enzymes are not consumed by the reactions they catalyse, nor do they alter the equilibrium of these reactions.

However, enzymes do differ from most other catalysts by being much more specific. Enzymes are known to catalyse about 4,000 biochemical reactions. A few RNA molecules called ribozymes also catalyse reactions, with an important example being some parts of the ribosome. Synthetic molecules called artificial enzymes also display enzyme-like catalysis.

Enzyme activity can be affected by other molecules. Inhibitors are molecules that decrease enzyme activity; activators are molecules that increase activity. Many drugs and poisons are enzyme inhibitors. Activity is also affected by temperature, chemical environment (*e.g.*, pH), and the concentration of substrate. Some enzymes are used commercially, for example, in the synthesis of antibiotics. In addition, some household products use enzymes to speed up biochemical reactions (*e.g.*, enzymes in biological washing powders break down protein or fat stains on clothes; enzymes in meat tenderizers break down proteins, making the meat easier to chew).

Structures and Mechanisms

Enzymes are generally globular proteins and range from just 62 amino acid residues in size, for the monomer of 4-oxalocrotonate tautomerase, to over 2,500 residues in the animal fatty acid synthase. A small number of RNA-based biological catalysts exist, with the most common being the ribosome; these are referred to as either RNA-enzymes or ribozymes. The activities of enzymes are determined by their three-dimensional structure. However, although structure does determine function, predicting a novel enzyme's activity just from its structure is a very difficult problem that has not yet been solved.

Most enzymes are much larger than the substrates they act on, and only a small portion of the enzyme (around 3–4 amino acids) is directly involved in catalysis. The region that contains these catalytic residues, binds the substrate, and then carries out the reaction is known as the active site. Enzymes can also contain sites that bind cofactors, which are needed for catalysis. Some enzymes also have binding sites for small molecules, which are often direct or indirect products or substrates of the reaction catalysed. This binding can serve to increase or decrease the enzyme's activity, providing a means for feedback regulation.

Like all proteins, enzymes are long, linear chains of amino acids that fold to produce a three-dimensional product. Each unique amino acid sequence produces a specific structure, which has unique properties. Individual protein chains may sometimes group together to form a protein complex.

Most enzymes can be denatured—that is, unfolded and inactivated—by heating or chemical denaturants, which disrupt the three-dimensional structure of the protein. Depending on the enzyme, denaturation may be reversible or irreversible. Structures of enzymes in complex with substrates or substrate analogs during a reaction may be obtained using Time resolved crystallography methods.

Specificity

Enzymes are usually very specific as to which reactions they catalyse and the substrates that are involved in these reactions. Complementary shape, charge and hydrophilic/hydrophobic characteristics of enzymes and substrates are responsible for this specificity. Enzymes can also show impressive levels of stereospecificity, regioselectivity and chemoselectivity.

Some of the enzymes showing the highest specificity and accuracy are involved in the copying and expression of the genome. These enzymes have "proof-reading" mechanisms. Here, an enzyme such as DNA polymerase catalyzes a reaction in a first step and then checks that the product is correct in a second step. This two-step process results in average error rates of less than 1 error in 100 million reactions in high-fidelity mammalian polymerases. Similar proofreading mechanisms are also found in RNA polymerase, aminoacyl tRNA synthetases and ribosomes. Some enzymes that produce secondary metabolites are described as promiscuous, as they can act on a relatively broad range of different substrates. It has been suggested that this broad substrate specificity is important for the evolution of new biosynthetic pathways.

"Lock and Key" Model

Enzymes are very specific, and it was suggested by the Nobel laureate organic chemist Emil Fischer in 1894 that this was because both the enzyme and the substrate possess specific complementary geometric shapes that fit exactly into one another. This is often referred to as "the lock and key" model. However, while this model explains enzyme specificity, it fails to explain the stabilization of the transition state that enzymes achieve. In 1958, Daniel Koshland suggested a modification to the lock and key model: since enzymes are rather flexible structures, the active site is continually reshaped by interactions with the substrate as the substrate interacts with the enzyme. As a result, the substrate does not simply bind to a rigid active site; the amino acid side chains which make up the active site are molded into the precise positions that enable the enzyme to perform its catalytic function.

In some cases, such as glycosidases, the substrate molecule also changes shape slightly as it enters the active site. The active site continues to change until the substrate is completely bound, at which point the final shape and charge is determined. Induced fit may enhance the fidelity of molecular recognition in the presence of competition and noise via the conformational proofreading mechanism.

Mechanisms

Enzymes can act in several ways, all of which lower $\Delta G^{\ddagger}$:

- Lowering the activation energy by creating an environment in which the transition state is stabilized (e.g. straining the shape of a substrate—by binding the transition-state conformation of the substrate/product molecules, the enzyme distorts the bound substrate(s) into their transition state form, thereby reducing the amount of energy required to complete the transition).
- Lowering the energy of the transition state, but without distorting the substrate, by creating an environment with the opposite charge distribution to that of the transition state.
- Providing an alternative pathway. For example, temporarily reacting with the substrate to form an intermediate ES complex, which would be impossible in the absence of the enzyme.
- Reducing the reaction entropy change by bringing substrates together in the correct orientation to react. Considering $\Delta H^{\ddagger}$ alone overlooks this effect.
- Increases in temperatures speed up reactions. Thus, temperature increases help the enzyme function and develop the end product even faster. However, if heated too much, the enzyme's shape deteriorates and only when the temperature comes back to normal does the enzyme regain its shape. Some enzymes like thermolabile enzymes work best at low temperatures.

Interestingly, this entropic effect involves destabilization of the ground state, and its contribution to catalysis is relatively small.

Transition State Stabilization

The understanding of the origin of the reduction of $\Delta G^{\ddagger}$ requires one to find out how the enzymes can stabilize its transition state more than the transition state of the uncatalyzed reaction. Apparently, the most effective way for reaching large stabilization is the use of electrostatic effects, in particular, by having a relatively fixed polar environment that is oriented toward the charge distribution of the transition state. Such an environment does not exist in the uncatalyzed reaction in water.

Dynamics and Function

The internal dynamics of enzymes is linked to their mechanism of catalysis. Internal dynamics are the movement of parts of the enzyme's structure, such as individual amino acid residues, a group of amino acids, or even an entire protein domain. These movements occur at various time-scales ranging from femtoseconds to seconds. Networks of protein residues throughout an enzyme's structure can contribute to catalysis through dynamic motions. Protein motions are vital to many enzymes, but whether small and fast vibrations, or larger and slower conformational movements are more important depends on the type of reaction involved. However, although these movements are important in binding and releasing substrates and products, it is not clear if protein movements help to accelerate the chemical steps in enzymatic reactions. These new insights also have implications in understanding allosteric effects and developing new drugs.

Allosteric Modulation

Allosteric sites are sites on the enzyme that bind to molecules in the cellular environment. The sites form weak, noncovalent bonds with these molecules, causing a change in the conformation of the enzyme. This change in conformation translates to the active site, which then affects the reaction rate of the enzyme. Allosteric interactions can both inhibit and activate enzymes and are a common way that enzymes are controlled in the body.

Cofactors and Coenzymes

Cofactors

Some enzymes do not need any additional components to show full activity. However, others require non-protein molecules called cofactors to be bound for activity. Cofactors can be either inorganic (*e.g.*, metal ions and iron-sulphur clusters) or organic compounds (e.g., flavin and heme). Organic cofactors can be either prosthetic groups, which are tightly bound to an enzyme, or coenzymes, which are released from the enzyme's active site during the reaction. Coenzymes include NADH, NADPH and adenosine triphosphate. These molecules transfer chemical groups between enzymes.

An example of an enzyme that contains a cofactor is carbonic anhydrase, and is shown in the ribbon diagram above with a zinc cofactor bound as part of its active site. These tightly bound molecules are usually found in the active site and are involved in catalysis. For example, flavin and heme cofactors are often involved in redox reactions.

Enzymes that require a cofactor but do not have one bound are called *apoenzymes* or *apoproteins*. An apoenzyme together with its cofactor(s) is called a *holoenzyme* (this is the active form). Most cofactors are not covalently attached to an enzyme, but are very tightly bound. However, organic prosthetic groups can be covalently bound (*e.g.*, thiamine pyrophosphate in the enzyme pyruvate dehydrogenase).

The term "holoenzyme" can also be applied to enzymes that contain multiple protein subunits, such as the DNA polymerases; here the holoenzyme is the complete complex containing all the subunits needed for activity.

Coenzymes

Coenzymes are small organic molecules that transport chemical groups from one enzyme to another. Some of these chemicals such as riboflavin, thiamine and folic acid are vitamins (compounds which cannot be synthesized by the body and must be acquired from the diet). The chemical groups carried include the hydride ion (H^-) carried by NAD or $NADP^+$, the phosphate group carried by adenosine triphosphate, the acetyl group carried by coenzyme A, formyl, methenyl or methyl groups carried by folic acid and the methyl group carried by S-adenosylmethionine. Since coenzymes are chemically changed as a consequence of enzyme action, it is useful to consider coenzymes to be a special class of substrates, or second substrates, which are common to many different enzymes. For example, about 700 enzymes are known to use the coenzyme NADH. Coenzymes are usually continuously regenerated and their concentrations maintained at a steady level inside the cell: for example, NADPH is regenerated through the pentose phosphate pathway and *S*-adenosylmethionine by methionine adenosyltransferase. This continuous regeneration means that even small amounts of coenzymes are used very intensively. For example, the human body turns over its own weight in ATP each day.

Thermodynamics

As all catalysts, enzymes do not alter the position of the chemical equilibrium of the reaction. Usually, in the presence of an enzyme, the reaction runs in the same direction as it would without the enzyme, just more quickly. However, in the absence of the enzyme, other possible uncatalyzed, "spontaneous" reactions might lead to different products, because in those conditions this different product is formed faster. Furthermore, enzymes can couple two or more reactions, so that a thermodynamically favourable reaction can be used to "drive" a thermodynamically unfavourable one. For example, the hydrolysis of ATP is often used to drive other chemical reactions. Enzymes catalyse the forward and backward reactions equally. They do not alter the equilibrium itself, but only the speed at which it is reached. For example, carbonic anhydrase catalyzes its reaction in either direction depending on the concentration of its reactants.

$$CO_2 + H_2O \xrightarrow{\text{Carbonic anhydrase}} H_2CO_3$$

(in tissues; high CO_2 concentration)

$$H_2CO_3 \xrightarrow{\text{Carbonic anhydrase}} CO_2 + H_2O$$

(in lungs; low CO_2 concentration)

Nevertheless, if the equilibrium is greatly displaced in one direction, that is, in a very exergonic reaction, the reaction is *effectively* irreversible. Under these conditions the enzyme will, in fact, only catalyse the reaction in the thermodynamically allowed direction.

Kinetics

Enzyme kinetics is the investigation of how enzymes bind substrates and turn them into products. The rate data used in kinetic analyses are obtained from enzyme assays.

In 1902 Victor Henri proposed a quantitative theory of enzyme kinetics, but his experimental data were not useful because the significance of the hydrogen ion concentration was not yet appreciated. After Peter Lauritz Sørensen had defined the logarithmic pH-scale and introduced the concept of buffering in 1909 the German chemist Leonor Michaelis and his Canadian postdoc Maud Leonora Menten repeated Henri's experiments and confirmed his equation which is referred to as Henri-Michaelis-Menten kinetics (sometimes also Michaelis-Menten kinetics).

Their work was further developed by G. E. Briggs and J. B. S. Haldane, who derived kinetic equations that are still widely used today. The major contribution of Henri was to think of enzyme reactions in two stages. In the first, the substrate binds reversibly to the enzyme, forming the enzyme-substrate complex. This is sometimes called the Michaelis complex. The enzyme then catalyzes the chemical step in the reaction and releases the product.

Enzymes can catalyse up to several million reactions per second. For example, the uncatalyzed decarboxylation of orotidine 5'-monophosphate has a half life of 78 million years. However, when the enzyme orotidine 5'-phosphate decarboxylase is added, the same process takes just 25 milliseconds. Enzyme rates depend on solution conditions and substrate concentration. Conditions that denature the protein abolish enzyme activity, such as high temperatures, extremes of pH or high salt concentrations, while raising substrate concentration tends to increase activity. To find the maximum speed of an enzymatic reaction, the substrate concentration is increased until a constant rate of product formation is seen.

This is shown in the saturation curve on the right. Saturation happens because, as substrate concentration increases, more and more of the free enzyme is converted into the substrate-bound ES form. At the maximum velocity (V_{max}) of the enzyme, all the enzyme active sites are bound to substrate, and the amount of ES complex is the same as the total amount of enzyme. However, V_{max} is only one kinetic constant of enzymes. The amount of substrate needed to achieve a given rate of reaction is also important. This is given by the Michaelis-Menten constant (K_m), which is the substrate concentration required for an enzyme to reach one-half its maximum velocity. Each enzyme has a characteristic K_m for a given substrate, and this can show how tight the binding of the substrate is to the enzyme. Another useful constant is k_{cat}, which is the number of substrate molecules handled by one active site per second.

The efficiency of an enzyme can be expressed in terms of k_{cat}/K_m. This is also called the specificity constant and incorporates the rate constants for all steps in the reaction. Because the specificity constant reflects both affinity and catalytic ability, it is useful for comparing different enzymes against each other, or the same enzyme with different substrates. The theoretical maximum for the specificity constant is called the diffusion limit and is about 10^8 to 10^9 (M^{-1} s^{-1}). At this point every collision of the enzyme with its substrate will result in catalysis, and the rate of product formation is not limited by the reaction rate but by the diffusion rate. Enzymes with this property are called *catalytically perfect* or *kinetically perfect*. Example of such enzymes are triose-phosphate isomerase, carbonic anhydrase, acetylcholinesterase, catalase, fumarase, β-lactamase, and superoxide dismutase.

Michaelis-Menten kinetics relies on the law of mass action, which is derived from the assumptions of free diffusion and thermodynamically driven random collision. However, many biochemical or cellular processes deviate significantly from these conditions, because of macromolecular crowding, phase-separation of the enzyme/substrate/product, or one or two-dimensional molecular movement. In these situations, a fractal Michaelis-Menten kinetics may be applied.

Some enzymes operate with kinetics which are faster than diffusion rates, which would seem to be impossible. Several mechanisms have been invoked to explain this phenomenon. Some proteins are believed to accelerate catalysis by drawing their substrate in and pre-orienting them by using dipolar electric fields. Other models invoke a quantum-mechanical tunnelling explanation, whereby a proton or an electron can tunnel through activation barriers, although for proton tunnelling this model remains somewhat controversial. Quantum tunnelling for protons has been observed in tryptamine. This suggests that enzyme catalysis may be more accurately characterized as "through the barrier" rather than the traditional model, which requires substrates to go "over" a lowered energy barrier.

Inhibition

Enzyme inhibitors are molecules that bind to enzymes and decrease their activity. Since blocking an enzyme's activity can kill a pathogen or correct a metabolic imbalance, many drugs are enzyme inhibitors. They are also used as herbicides and pesticides. Not all molecules that bind to enzymes are inhibitors; *enzyme activators* bind to enzymes and increase their enzymatic activity.

The binding of an inhibitor can stop a substrate from entering the enzyme's active site and/or hinder the enzyme from catalysing its reaction. Inhibitor binding is either reversible or irreversible. Irreversible inhibitors usually react with the enzyme and change it chemically. These inhibitors modify key amino acid residues needed for enzymatic activity. In contrast, reversible inhibitors bind non-covalently and different types of inhibition are produced depending on whether these inhibitors bind the enzyme, the enzyme-substrate complex, or both.

Many drug molecules are enzyme inhibitors, so their discovery and improvement is an active area of research in biochemistry and pharmacology. A medicinal enzyme inhibitor is often judged by its specificity (its lack of binding to other proteins) and its potency (its dissociation constant, which indicates the concentration needed to inhibit the enzyme). A high specificity and potency ensure that a drug will have few side effects and thus low toxicity.

Enzyme inhibitors also occur naturally and are involved in the regulation of metabolism. For example, enzymes in a metabolic pathway can be inhibited by downstream products. This type of negative feedback slows flux through a pathway when the products begin to build up and is an important way to maintain homeostasis in a cell. Other cellular enzyme inhibitors are proteins that specifically bind to and inhibit an enzyme target. This can help control enzymes that may be damaging to a cell, such as proteases or nucleases; a well-characterised example is the ribonuclease inhibitor, which binds to ribonucleases in one of the tightest known protein–protein interactions. Natural enzyme inhibitors can also be poisons and are used as defences against predators or as ways of killing prey.

Reversible Inhibitors

Types of Reversible Inhibitors

Reversible inhibitors bind to enzymes with non-covalent interactions such as hydrogen bonds, hydrophobic interactions and ionic bonds. Multiple weak bonds between the inhibitor and the active site combine to produce strong and specific binding. In contrast to substrates and irreversible inhibitors, reversible inhibitors generally do not undergo chemical reactions when bound to the enzyme and can be easily removed by dilution or dialysis.

There are four kinds of reversible enzyme inhibitors. They are classified according to the effect of varying the concentration of the enzyme's substrate on the inhibitor.

- In competitive inhibition, the substrate and inhibitor cannot bind to the enzyme at the same time. This usually results from the inhibitor having an affinity for the active site of an enzyme where the substrate also binds; the substrate and inhibitor *compete* for access to the enzyme's active site. This type of inhibition can be overcome by sufficiently high concentrations of substrate, i.e., by out-competing the inhibitor. Competitive inhibitors are often similar in structure to the real substrate.
- In uncompetitive inhibition, the inhibitor binds only to the substrate-enzyme complex, it should not be confused with non-competitive inhibitors. Both Vmax and Km decrease (maximum velocity decreases as a result of removing activated complex while binding efficiency increases as a result of Le Chatelier's principle).
- In mixed inhibition, the inhibitor can bind to the enzyme at the same time as the enzyme's substrate. However, the binding of the inhibitor affects the binding of the substrate, and vice versa. This type of inhibition can be reduced, but not overcome by increasing concentrations of substrate. Although it is possible for mixed-type inhibitors to bind in the active site, this type of inhibition generally results from an allosteric effect where the inhibitor binds to a different site on an enzyme. Inhibitor binding to this allosteric site changes the conformation (i.e., tertiary structure or three-dimensional shape) of the enzyme so that the affinity of the substrate for the active site is reduced.
- Non-competitive inhibition is a form of mixed inhibition where the binding of the inhibitor to the enzyme reduces its activity but does not affect the binding of substrate. As a result, the extent of inhibition depends only on the concentration of the inhibitor.

Quantitative Description of Reversible Inhibition

Reversible inhibition can be described quantitatively in terms of the inhibitor's binding to the enzyme and to the enzyme–substrate complex, and its effects on the kinetic constants of the enzyme. In the classic Michaelis–Menten scheme below, an enzyme (E) binds to its substrate (S) to form the enzyme–substrate complex ES. Upon catalysis, this complex breaks down to release product P and free enzyme. The inhibitor (I) can bind to either E or ES with the dissociation constants K_i or K_i', respectively.

When an enzyme has multiple substrates, inhibitors can show different types of inhibition depending on which substrate is considered. This results from the active site containing two different binding sites within the active site, one for each substrate. For example, an inhibitor might compete with substrate A for the first binding site, but be a non-competitive inhibitor with respect to substrate B in the second binding site.

Measuring the Dissociation Constants of a Reversible Inhibitor

As noted above, an enzyme inhibitor is characterised by its two dissociation constants, K_i and K_i', to the enzyme and to the enzyme-substrate complex, respectively. The enzyme-inhibitor constant K_i can be measured directly by various methods; one extremely accurate method is isothermal titration calorimetry, in which the inhibitor is titrated into a solution of enzyme and the heat released or absorbed is measured.

However, the other dissociation constant K_i' is difficult to measure directly, since the enzyme-substrate complex is short-lived and undergoing a chemical reaction to form the product. Hence, K_i' is usually measured indirectly, by observing the enzyme activity under various substrate and inhibitor concentrations, and fitting the data to a modified Michaelis–Menten equation.

$$V = \frac{V_{max}[S]}{\alpha K_m + \alpha'[S]} = \frac{(1/\alpha')V_{max}[S]}{(\alpha/\alpha')K_m + [S]}$$

where the modifying factors α and α' are defined by the inhibitor concentration and its two dissociation constants

$$\alpha = 1 + \frac{[I]}{K_i}$$

$$\alpha' = 1 + \frac{[I]}{K_i'}.$$

Thus, in the presence of the inhibitor, the enzyme's effective K_m and V_{max} become $(\alpha/\alpha')K_m$ and $(1/\alpha')V_{max}$, respectively. However, the modified Michaelis-Menten equation assumes that binding of the inhibitor to the enzyme has reached equilibrium, which may be a very slow process for inhibitors with sub-nanomolar dissociation constants. In these cases, it is usually more practical to treat the tight-binding inhibitor as an irreversible inhibitor; however, it can still be possible to estimate K_i' kinetically if K_i is measured independently.

The effects of different types of reversible enzyme inhibitors on enzymatic activity can be visualized using graphical representations of the

Michaelis–Menten equation, such as Lineweaver–Burk and Eadie-Hofstee plots. For example, in the Lineweaver–Burk plots at the right, the competitive inhibition lines intersect on the *y*-axis, illustrating that such inhibitors do not affect V_{max}. Similarly, the non-competitive inhibition lines intersect on the *x*-axis, showing these inhibitors do not affect K_m. However, it can be difficult to estimate K_i and K_i' accurately from such plots, so it is advisable to estimate these constants using more reliable nonlinear regression methods, as described above.

Special Cases

- The mechanism of partially co mpetitive inhibition is similar to that of non-competitive, except that the EIS complex has catalytic activity, which may be lower or even higher (partially competitive activation) than that of the enzyme–substrate (ES) complex. This inhibition typically displays a lower V_{max}, but an unaffected K_m value.
- Uncompetitive inhibition occurs when the inhibitor binds only to the enzyme–substrate complex, not to the free enzyme; the EIS complex is catalytically inactive. This mode of inhibition is rare and causes a decrease in both V_{max} and the K_m value.
- Substrate and product inhibition is where either the substrate or product of an enzyme reaction inhibit the enzyme's activity. This inhibition may follow the competitive, uncompetitive or mixed patterns. In substrate inhibition there is a progressive decrease in activity at high substrate concentrations. This may indicate the existence of two substrate-binding sites in the enzyme. At low substrate, the high-affinity site is occupied and normal kinetics are followed. However, at higher concentrations, the second inhibitory site becomes occupied, inhibiting the enzyme. Product inhibition is often a regulatory feature in metabolism and can be a form of negative feedback.
- Slow-tight inhibition occurs when the initial enzyme–inhibitor complex EI undergoes isomerisation to a second more tightly held complex, EI*, but the overall inhibition process is reversible. This manifests itself as slowly increasing enzyme inhibition. Under these conditions, traditional Michaelis–Menten kinetics give a false value for K_i, which is time–dependent. The true value of K_i can be obtained through more complex analysis of the on (k_{on}) and off (k_{off}) rate constants for inhibitor association.

Examples of Reversible Inhibitors

As enzymes have evolved to bind their substrates tightly, and most reversible inhibitors bind in the active site of enzymes, it is unsurprising that some of these inhibitors are strikingly similar in structure to the substrates of their targets. An example of these substrate mimics are the protease inhibitors, a very successful class of antiretroviral drugs used to treat HIV. The structure of ritonavir, a protease inhibitor based on a peptide and containing three peptide bonds, is shown on the right. As this drug resembles the protein that is the substrate of the HIV protease, it competes with this substrate in the enzyme's active site.

Enzyme inhibitors are often designed to mimic the transition state or intermediate of an enzyme-catalysed reaction. This ensures that the inhibitor exploits the transition state stabilising effect of the enzyme, resulting in a better binding affinity (lower K_i) than substrate-based designs. An example of such a transition state inhibitor is the antiviral drug oseltamivir; this drug mimics the planar nature of the ring oxonium ion in the reaction of the viral enzyme neuraminidase.

However, not all inhibitors are based on the structures of substrates. For example, the structure of another HIV protease inhibitor tipranavir is shown on the left. This molecule is not based on a peptide and has no obvious structural similarity to a protein substrate. These non-peptide inhibitors can be more stable than inhibitors containing peptide bonds, because they will not be substrates for peptidases and are less likely to be degraded.

In drug design it is important to consider the concentrations of substrates to which the target enzymes are exposed. For example, some protein kinase inhibitors have chemical structures that are similar to adenosine triphosphate, one of the substrates of these enzymes.

However, drugs that are simple competitive inhibitors will have to compete with the high concentrations of ATP in the cell. Protein kinases can also be inhibited by competition at the binding sites where the kinases interact with their substrate proteins, and most proteins are present inside cells at concentrations much lower than the concentration of ATP. As a consequence, if two protein kinase inhibitors both bind in the active site with similar affinity, but only one has to compete with ATP, then the competitive inhibitor at the protein-binding site will inhibit the enzyme more effectively.

Irreversible Inhibitors

Types of Irreversible Inhibition

Irreversible inhibitors usually covalently modify an enzyme, and inhibition cannot therefore be reversed. Irreversible inhibitors often contain reactive functional groups such as nitrogen mustards, aldehydes, haloalkanes, alkenes, Michael acceptors, phenyl sulfonates, or fluorophosphonates. These electrophilic groups react with amino acid side chains to form covalent adducts. The residues modified are those with side chains containing nucleophiles such as hydroxyl or sulfhydryl groups; these include the amino acids serine (as in DFP, right), cysteine, threonine or tyrosine.

Irreversible inhibition is different from irreversible enzyme inactivation. Irreversible inhibitors are generally specific for one class of enzyme and do not inactivate all proteins; they do not function by destroying protein structure but by specifically altering the active site of their target. For example, extremes of pH or temperature usually cause denaturation of all protein structure, but this is a non-specific effect. Similarly, some non-specific chemical treatments destroy protein structure: for example, heating in concentrated hydrochloric acid will hydrolyse the peptide bonds holding proteins together, releasing free amino acids.

Irreversible inhibitors display time-dependent inhibition and their potency therefore cannot be characterised by an IC_{50} value. This is because the amount of active enzyme at a given concentration of irreversible inhibitor will be different depending on how long the inhibitor is pre-incubated with the enzyme. Instead, $k_{obs}/[I]$ values are used, wherek_{obs} is the observed pseudo-first order rate of inactivation (obtained by plotting the log of % activity vs. time) and $[I]$ is the concentration of inhibitor. The $k_{obs}/[I]$ parameter is valid as long as the inhibitor does not saturate binding with the enzyme (in which case $k_{obs} = k_{inact}$).

Analysis of Irreversible Inhibition

The irreversible inhibitors form a reversible non-covalent complex with the enzyme (EI or ESI) and this then reacts to produce the covalently modified "dead-end complex" EI*. The rate at which EI* is formed is called the inactivation rate or k_{inact}. Since formation of EI may compete with ES, binding of irreversible inhibitors can be prevented by competition either with substrate or with a second, reversible inhibitor. This protection effect is good evidence of a specific reaction of the irreversible inhibitor with the active site.

The binding and inactivation steps of this reaction are investigated by incubating the enzyme with inhibitor and assaying the amount of activity remaining over time. The activity will be decrease in a time-dependent manner, usually following exponential decay. Fitting these data to a rate equation gives the rate of inactivation at this concentration of inhibitor.

This is done at several different concentrations of inhibitor. If a reversible EI complex is involved the inactivation rate will be saturable and fitting this curve will give k_{inact} and K_i. Another method that is widely used in these analyses is mass spectrometry. Here, accurate measurement of the mass of the unmodified native enzyme and the inactivated enzyme gives the increase in mass caused by reaction with the inhibitor and shows the stoichiometry of the reaction.

This is usually done using a MALDI-TOF mass spectrometer. In a complementary technique, peptide mass fingerprinting involves digestion of the native and modified protein with a protease such as trypsin. This will produce a set of peptides that can be analysed using a mass spectrometer. The peptide that changes in mass after reaction with the inhibitor will be the one that contains the site of modification.

Special Cases

Not all irreversible inhibitors form covalent adducts with their enzyme targets. Some reversible inhibitors bind so tightly to their target enzyme that they are essentially irreversible. These tight-binding inhibitors may show kinetics similar to covalent irreversible inhibitors. In these cases, some of these inhibitors rapidly bind to the enzyme in a low-affinity EI complex and this then undergoes a slower rearrangement to a very tightly bound EI* complex. This kinetic behaviour is called slow-binding. This slow rearrangement after binding often involves a conformational change as the enzyme "clamps down" around the inhibitor molecule. Examples of slow-binding inhibitors include some important drugs, such methotrexate, allopurinol, and the activated form of acyclovir.

Examples of Irreversible Inhibitors

The enzyme hydrolyses the phosphorus–fluorine bond, but the phosphate residue remains bound to the serine in the active site, deactivating it. Similarly, DFP also reacts with the active site of acetylcholine esterase in the synapses of neurons, and consequently is a potent neurotoxin, with a lethal dose of less than 100 mg.

Suicide inhibition is an unusual type of irreversible inhibition where the enzyme converts the inhibitor into a reactive form in its active site. An example is the inhibitor of polyamine biosynthesis, α-difluoromethylornithine or DFMO, which is an analogue of the amino acid ornithine, and is used to treat African trypanosomiasis (sleeping sickness). Ornithine decarboxylase can catalyse the decarboxylation of DFMO instead of ornithine, as shown above. However, this decarboxylation reaction is followed by the elimination of a fluorine atom, which converts this catalytic intermediate into a conjugated imine, a highly electrophilic species. This reactive form of DFMO then reacts with either a cysteine or lysine residue in the active site to irreversibly inactivate the enzyme.

Discovery and Design of Inhibitors

New drugs are the products of a long drug development process, the first step of which is often the discovery of a new enzyme inhibitor. In the past the only way to discover these new inhibitors was by trial and error: screening huge libraries of compounds against a target enzyme and hoping that some useful leads would emerge. This brute force approach is still successful and has even been extended by combinatorial chemistry approaches that quickly produce large numbers of novel compounds and high-throughput screening technology to rapidly screen these huge chemical libraries for useful inhibitors.

More recently, an alternative approach has been applied: rational drug design uses the three-dimensional structure of an enzyme's active site to predict which molecules might be inhibitors. These predictions are then tested and one of these tested compounds may be a novel inhibitor. This new inhibitor is then used to try to obtain a structure of the enzyme in an inhibitor/enzyme complex to show how the molecule is binding to the active site, allowing changes to be made to the inhibitor to try to optimise binding. This test and improve cycle is then repeated until a sufficiently potent inhibitor is produced. Computer-based methods of predicting the affinity of an inhibitor for an enzyme are also being developed, such as molecular docking and molecular mechanics.

Uses of Inhibitors

Enzyme inhibitors are found in nature and are also designed and produced as part of pharmacology and biochemistry. Natural poisons are often enzyme inhibitors that have evolved to defend a plant or animal against predators. These natural toxins include some of the most poisonous compounds known. Artificial inhibitors are often used as drugs, but can also be insecticides such as malathion, herbicides such as glyphosate, or disinfectants such as triclosan.

Chemotherapy

The most common uses for enzyme inhibitors are as drugs to treat disease. Many of these inhibitors target a human enzyme and aim to correct a pathological condition. However, not all drugs are enzyme inhibitors. Some, such as anti-epileptic drugs, alter enzyme activity by causing more or less of the enzyme to be produced. These effects are called enzyme induction and inhibition and are alterations in gene expression, which is unrelated to the type of enzyme inhibition. Other drugs interact with cellular targets that are not enzymes, such as ion channels or membrane receptors.

An example of a medicinal enzyme inhibitor is sildenafil (Viagra), a common treatment for male erectile dysfunction. This compound is a potent inhibitor of cGMP specific phosphodiesterase type 5, the enzyme that degrades the signalling molecule cyclic guanosine monophosphate. This signalling molecule triggers smooth muscle relaxation and allows blood flow into the corpus cavernosum, which causes an erection. Since the drug decreases the activity of the enzyme that halts the signal, it makes this signal last for a longer period of time.

Folic acid is a substrate of dihydrofolate reductase, an enzyme involved in making nucleotides that is potently inhibited by methotrexate. Methotrexate blocks the action of dihydrofolate reductase and thereby halts the production of nucleotides. This block of nucleotide biosynthesis is more toxic to rapidly growing cells than non-dividing cells, since a rapidly-growing cell has to carry out DNA replication, therefore methotrexate is often used in cancer chemotherapy.

Drugs also are used to inhibit enzymes needed for the survival of pathogens. For example, bacteria are surrounded by a thick cell wall made of a net-like polymer called peptidoglycan. Many antibiotics such as penicillin and vancomycin inhibit the enzymes that produce and then cross-link the strands of this polymer together. This causes the cell wall to lose strength and the bacteria to burst.

Drug design is facilitated when an enzyme that is essential to the pathogen's survival is absent or very different in humans. In the example above, humans do not make peptidoglycan, therefore inhibitors of this process are selectively toxic to bacteria. Selective toxicity is also produced in antibiotics by exploiting differences in the structure of the ribosomes in bacteria, or how they make fatty acids.

Metabolic Control

Enzyme inhibitors are also important in metabolic control. Many metabolic pathways in the cell are inhibited by metabolites that control enzyme activity through allosteric regulation or substrate inhibition. A good example is the allosteric regulation of the glycolytic pathway. This catabolic pathway consumes glucose and produces ATP, NADH and pyruvate. A key step for the regulation of glycolysis is an early reaction in the pathway catalysed by phosphofructokinase-1 (PFK1). When ATP levels rise, ATP binds an allosteric site in PFK1 to decrease the rate of the enzyme reaction; glycolysis is inhibited and ATP production falls.

This negative feedback control helps maintain a steady concentration of ATP in the cell. However, metabolic pathways are not just regulated through inhibition since enzyme activation is equally important. With respect to PFK1, fructose 2,6-bisphosphate and ADP are examples of metabolites that are allosteric activators.

Physiological enzyme inhibition can also be produced by specific protein inhibitors. This mechanism occurs in the pancreas, which synthesises many digestive precursor enzymes known as zymogens. Many of these are activated by the trypsin protease, so it is important to inhibit the activity of trypsin in the pancreas to prevent the organ from digesting itself. One way in which the activity of trypsin is controlled is the production of a specific and potent trypsin inhibitor protein in the pancreas. This inhibitor binds tightly to trypsin, preventing the trypsin activity that would otherwise be detrimental to the organ. Although the trypsin inhibitor is a protein, it avoids being hydrolysed as a substrate by the protease by excluding water from trypsin's active site and destabilising the transition state. Other examples of physiological enzyme inhibitor proteins include the barstar inhibitor of the bacterial ribonuclease barnase and the inhibitors of protein phosphatases.

Pesticides and Herbicides

Many herbicides and pesticides are enzyme inhibitors. Acetylcholinesterase (AChE) is an enzyme found in animals from insects to humans. It is essential to nerve cell function through its mechanism of breaking down the neurotransmitter acetylcholine into its constituents, acetate and choline. This is somewhat unique among neurotransmitters as most, including serotonin, dopamine, and norepinephrine, are absorbed from the synaptic cleft rather than cleaved.

A large number of AChE inhibitors are used in both medicine and agriculture. Reversible competitive inhibitors, such as edrophonium, physostigmine, and neostigmine, are used in the treatment of myasthenia gravis and in anaesthesia. The carbamate pesticides are also examples of reversible AChE inhibitors. The organophosphate insecticides such as malathion, parathion, and chlorpyrifos irreversibly inhibit acetylcholinesterase.

The herbicide glyphosate is an inhibitor of 3-phosphoshikimate 1-carboxyvinyltransferase, other herbicides, such as the sulfonylureas inhibit the enzyme acetolactate synthase. Both these enzymes are needed for plants to make branched-chain amino acids. Many other enzymess are inhibited by herbicides, including enzymes needed for the biosynthesis of lipids and carotenoids and the processes of photosynthesis and oxidative phosphorylation.

Natural Poisons

Animals and plants have evolved to synthesise a vast array of poisonous products including secondary metabolites, peptides and proteins that can act as inhibitors. Natural toxins are usually small organic molecules and are so diverse that there are probably natural inhibitors for most metabolic processes. The metabolic processes targeted by natural poisons encompass more than enzymes in metabolic pathways and can also include the inhibition of receptor, channel and structural protein functions in a cell. For example, paclitaxel (taxol), an organic molecule found in the Pacific yew tree, binds tightly to tubulin dimers and inhibits their assembly into microtubules in the cytoskeleton.

Many natural poisons act as neurotoxins that can cause paralysis leading to death and have functions for defence against predators or in hunting and capturing prey. Some of these natural inhibitors, despite their toxic attributes, are valuable for therapeutic uses at lower doses. An example of a neurotoxin are the glycoalkaloids, from the plant species in the *Solanaceae* family (includes potato, tomato and eggplant), that are acetylcholinesterase inhibitors. Inhibition of this enzyme causes an uncontrolled increase in the acetylcholine neurotransmitter, muscular paralysis and then death. Neurotoxicity can also result from the inhibition of receptors; for example, atropine from deadly nightshade (*Atropa belladonna*) that functions as a competitive antagonist of the muscarinic acetylcholine receptors.

Although many natural toxins are secondary metabolites, these poisons also include peptides and proteins. An example of a toxic peptide is alpha-amanitin, which is found in relatives of the death cap mushroom. This is a potent enzyme inhibitor, in this case preventing the RNA polymerase II enzyme from transcribing DNA.

The algal toxin microcystin is also a peptide and is an inhibitor of protein phosphatases. This toxin can contaminate water supplies after algal blooms and is a known carcinogen that can also cause acute liver hemorrhage and death at higher doses. Proteins can also be natural poisons or antinutrients, such as the trypsin inhibitors that are found in some legumes. A less common class of toxins are toxic enzymes: These act as irreversible inhibitors of their target enzymes and work by chemically modifying their substrate enzymes.

An example is ricin, an extremely potent protein toxin found in castor oil beans. This enzyme is a glycosidase that inactivates ribosomes. Since ricin is a catalytic irreversible inhibitor, this allows just a single molecule of ricin to kill a cell. Enzyme reaction rates can be decreased by various types of enzyme inhibitors.

Competitive Inhibition

In competitive inhibition, the inhibitor and substrate compete for the enzyme (i.e., they can not bind at the same time). Often competitive inhibitors strongly resemble the real substrate of the enzyme. For example, methotrexate is a competitive inhibitor of the enzyme dihydrofolate reductase, which catalyzes the reduction of dihydrofolate to tetrahydrofolate.

The similarity between the structures of folic acid and this drug. Note that binding of the inhibitor need *not* be to the substrate binding site (as frequently stated), if binding of the inhibitor changes the conformation of the enzyme to prevent substrate binding and *vice versa*. In competitive inhibition the maximal velocity of the reaction is not changed, but higher substrate concentrations are required to reach a given velocity, increasing the apparent K_m.

Uncompetitive Inhibition

In uncompetitive inhibition the inhibitor can not bind to the free enzyme, but only to the ES-complex. The EIS-complex thus formed is enzymatically inactive. This type of inhibition is rare, but may occur in multimeric enzymes.

Non-competitive Inhibition

Non-competitive inhibitors can bind to the enzyme at the same time as the substrate, i.e. they *never* bind to the active site. Both the EI and EIS complexes are enzymatically inactive. Because the inhibitor can not be driven from the enzyme by higher substrate concentration (in contrast to competitive inhibition), the apparent V_{max} changes. But because the substrate can still bind to the enzyme, the K_m stays the same.

Mixed Inhibition

This type of inhibition resembles the non-competitive, except that the EIS-complex has residual enzymatic activity.

In many organisms inhibitors may act as part of a feedback mechanism. If an enzyme produces too much of one substance in the organism, that substance may act as an inhibitor for the enzyme at the beginning of the pathway that produces it, causing production of the substance to slow down or stop when there is sufficient amount. This is a form of negative feedback. Enzymes which are subject to this form of regulation are often multimeric and have allosteric binding sites for regulatory substances. Their substrate/velocity plots are not hyperbolar, but sigmoidal (S-shaped).

Irreversible inhibitors react with the enzyme and form a covalent adduct with the protein. The inactivation is irreversible. These compounds include eflornithine a drug used to treat the parasitic disease sleeping sickness. Penicillin and Aspirin also act in this manner. With these drugs, the compound is bound in the active site and the enzyme then converts the inhibitor into an activated form that reacts irreversibly with one or more amino acid residues.

Uses of Inhibitors

Since inhibitors modulate the function of enzymes they are often used as drugs. An common example of an inhibitor that is used as a drug is aspirin, which inhibits the COX-1 and COX-2 enzymes that produce the inflammation messenger prostaglandin, thus suppressing pain and inflammation. However, other enzyme inhibitors are poisons. For example, the poison cyanide is an irreversible enzyme inhibitor that combines with the copper and iron in the active site of the enzyme cytochrome c oxidase and blocks cellular respiration.

Biological Function

Enzymes serve a wide variety of functions inside living organisms. They are indispensable for signal transduction and cell regulation, often via kinases and phosphatases. They also generate movement, with myosin hydrolysing ATP to generate muscle contraction and also moving cargo around the cell as part of the cytoskeleton. Other ATPases in the cell membrane are ion pumps involved in active transport. Enzymes are also involved in more exotic functions, such as luciferase generating light in fireflies. Viruses can also contain enzymes for infecting cells, such as the HIV integrase and reverse transcriptase, or for viral release from cells, like the influenza virus neuraminidase.

An important function of enzymes is in the digestive systems of animals. Enzymes such as amylases and proteases break down large molecules (starch or proteins, respectively) into smaller ones, so they can be absorbed by the intestines. Starch molecules, for example, are too large to be absorbed from the intestine, but enzymes hydrolyse the starch chains into smaller molecules such as maltose and eventually glucose, which can then be absorbed. Different enzymes digest different food substances. In ruminants which have herbivorous diets, microorganisms in the gut produce another enzyme, cellulase to break down the cellulose cell walls of plant fiber.

Several enzymes can work together in a specific order, creating metabolic pathways. In a metabolic pathway, one enzyme takes the product of another enzyme as a substrate. After the catalytic reaction, the product is then passed on to another enzyme. Sometimes more than one enzyme can catalyse the same reaction in parallel, this can allow more complex regulation: with for example a low constant activity being provided by one enzyme but an inducible high activity from a second enzyme.

Enzymes determine what steps occur in these pathways. Without enzymes, metabolism would neither progress through the same steps, nor be fast enough to serve the needs of the cell. Indeed, a metabolic pathway such as glycolysis could not exist independently of enzymes. Glucose, for example, can react directly with ATP to become phosphorylated at one or more of its carbons. In the absence of enzymes, this occurs so slowly as to be insignificant. However, if hexokinase is added, these slow reactions continue to take place except that phosphorylation at carbon 6 occurs so rapidly that if the mixture is tested a short time later, glucose-6-phosphate is found to be the only significant product. Consequently, the network of metabolic pathways within each cell depends on the set of functional enzymes that are present.

Control of Activity

There are five main ways that enzyme activity is controlled in the cell.

1. Enzyme production (transcription and translation of enzyme genes) can be enhanced or diminished by a cell in response to changes in the cell's environment. This form of gene regulation is called enzyme induction and inhibition. For example, bacteria may become resistant to antibiotics such as penicillin because enzymes called beta-lactamases are induced that hydrolyse the crucial beta-lactam ring within the penicillin molecule. Another example are enzymes in the liver called cytochrome P450 oxidases, which are important in drug metabolism. Induction or inhibition of these enzymes can cause drug interactions.
2. Enzymes can be compartmentalized, with different metabolic pathways occurring in different cellular compartments. For example, fatty acids are synthesized by one set of enzymes in the cytosol, endoplasmic reticulum and the Golgi apparatus and used by a different set of enzymes as a source of energy in the mitochondrion, through b-oxidation.
3. Enzymes can be regulated by inhibitors and activators. For example, the end product(s) of a metabolic pathway are often inhibitors for one of the first enzymes of the pathway (usually the first irreversible step, called *committed step*), thus regulating the amount of end product made by the pathways. Such a regulatory mechanism is called a negative feedback mechanism, because the amount of the end product produced is regulated by its own concentration. Negative feedback mechanism can effectively adjust the rate of synthesis of intermediate metabolites according to the demands of the cells.

This helps allocate materials and energy economically, and prevents the manufacture of excess end products. The control of enzymatic action helps to maintain a stable internal environment in living organisms.

4. Enzymes can be regulated through post-translational modification. This can include phosphorylation, myristoylation and glycosylation. For example, in the response to insulin, the phosphorylation of multiple enzymes, including glycogen synthase, helps control the synthesis or degradation of glycogen and allows the cell to respond to changes in blood sugar. Another example of post-translational modification is the cleavage of the polypeptide chain. Chymotrypsin, a digestive protease, is produced in inactive form as chymotrypsinogen in the pancreas and transported in this form to the stomach where it is activated. This stops the enzyme from digesting the pancreas or other tissues before it enters the gut. This type of inactive precursor to an enzyme is known as a zymogen.
5. Some enzymes may become activated when localized to a different environment (e.g. from a reducing (cytoplasm) to an oxidizing (periplasm) environment, high pH to low pH etc.). For example, hemagglutinin in the influenza virus is activated by a conformational change caused by the acidic conditions, these occur when it is taken up inside its host cell and enters the lysosome.

Involvement in Disease

Since the tight control of enzyme activity is essential for homeostasis, any malfunction (mutation, overproduction, underproduction or deletion) of a single critical enzyme can lead to a genetic disease. The importance of enzymes is shown by the fact that a lethal illness can be caused by the malfunction of just one type of enzyme out of the thousands of types present in our bodies.

One example is the most common type of phenylketonuria. A mutation of a single amino acid in the enzyme phenylalanine hydroxylase, which catalyzes the first step in the degradation of phenylalanine, results in build-up of phenylalanine and related products. This can lead to mental retardation if the disease is untreated.

Another example is when germline mutations in genes coding for DNA repair enzymes cause hereditary cancer syndromes such as xeroderma pigmentosum. Defects in these enzymes cause cancer since the body is less able to repair mutations in the genome. This causes a slow accumulation of mutations and results in the development of many types of cancer in the sufferer.

Naming Conventions

An enzyme's name is often derived from its substrate or the chemical reaction it catalyzes, with the word ending in *-ase*. Examples are lactase, alcohol dehydrogenase and DNA polymerase. This may result in different enzymes, called isozymes, with the same function having the same basic name. Isoenzymes have a different amino acid sequence and might be distinguished by their optimal pH, kinetic properties or immunologically. Furthermore, the normal physiological reaction an enzyme catalyzes may not be the same as under artificial conditions. This can result in the same enzyme being identified with two different names. *E.g.* Glucose isomerase, used industrially to convert glucose into the sweetener fructose, is a xylose isomerase *in vivo*.

The International Union of Biochemistry and Molecular Biology have developed a nomenclature for enzymes, the EC numbers; each enzyme is described by a sequence of four numbers preceded by "EC". The first number broadly classifies the enzyme based on its mechanism.

The top-level classification is:

- EC 1 *Oxidoreductases*: catalyse oxidation/reduction reactions
- EC 2 *Transferases*: transfer a functional group (*e.g.* a methyl or phosphate group)
- EC 3 *Hydrolases*: catalyse the hydrolysis of various bonds
- EC 4 *Lyases*: cleave various bonds by means other than hydrolysis and oxidation
- EC 5 *Isomerases*: catalyse isomerization changes within a single molecule
- EC 6 *Ligases*: join two molecules with covalent bonds.

According to the naming conventions, enzymes are generally classified into six main family classes and many sub-family classes. Some web-servers, e.g., EzyPred and bioinformatics tools have been developed to predict which main family class and sub-family class an enzyme molecule belongs to according to its sequence information alone via the pseudo amino acid composition.

Industrial Applications

Enzymes are used in the chemical industry and other industrial applications when extremely specific catalysts are required. However, enzymes in general are limited in the number of reactions they have evolved to catalyse and also by their lack of stability in organic solvents and at high temperatures. Consequently, protein engineering is an active area of research and involves attempts to create new enzymes with novel properties, either through rational design or *in vitro* evolution. These efforts have begun to be successful, and a few enzymes have now been desiged "from scratch" to catalyse reactions that do not occur in nature.

Bioenergetics

Bioenergetics is the part of biochemistry concerned with the energy involved in making and breaking of chemical bonds in the molecules found in biological organisms.

Growth, development and metabolism are some of the central phenomena in the study of biological organisms. The role of energy is fundamental to such biological processes. The ability to harness energy from a variety of metabolic pathways is a property of all living organisms. Life is dependent on energy transformations; living organisms survive because of exchange of energy within and without.

In a living organism, chemical bonds are broken and made as part of the exchange and transformation of energy. Energy is available for work (such as mechanical work) or for other processes (such as chemical synthesis and anabolic processes in growth), when weak bonds are broken and stronger bonds are made. The production of stronger bonds allows release of usable energy.

Living organisms obtain energy from organic and inorganic materials. For example, lithotrophs can oxidize minerals such as nitrates or forms of sulphur, such as elemental sulphur, sulfites, and hydrogen sulfide to produce ATP. In photosynthesis, autotrophs can produce ATP using light energy. Heterotrophs must consume organic compounds. These are mostly carbohydrates, fats, and proteins. The amount of energy actually obtained by the organism is lower than the amount present in the food; there are losses in digestion, metabolism, and thermogenesis.

The materials are generally combined with oxygen to release energy, although some can also be oxidized anaerobically by various organisms. The bonds holding the molecules of nutrients together and the bonds holding molecules of free oxygen together are all relatively weak compared with the chemical bonds holding carbon dioxide and water together. The utilization of these materials is a form of slow combustion. That is why the energy content of food can be estimated with a bomb calorimeter. The materials are oxidized slowly enough that the organisms do not actually produce fire. The oxidation releases energy because stronger bonds have been formed. This net energy may evolve as heat, or some of which may be used by the organism for other purposes, such as breaking other bonds to do chemistry.

Living organisms produce ATP from energy sources via oxidative phosphorylation. The terminal phosphate bonds of ATP are relatively weak compared with the stronger bonds formed when ATP is broken down to adenosine monophosphate and phosphate, dissolved in water. Here it is the energy of hydration that results in energy release. This hydrolysis of ATP is used as a battery to store energy in cells, for intermediate metabolism. Utilization of chemical energy from such molecular bond rearrangement powers biological processes in every biological organism.

Types of Reactions

- Exergonic is a spontaneous reaction that releases energy. It is thermodynamically favoured. On the course of a reaction, energy needs to be put in, this activation energy drives the reactants from a stable state to a highly energetic unstable configuration. These reactants are usually complex molecules that are broken into simpler products. The entire reaction is usually catabolic. The release of energy, also called free energy is a $-\Delta G$ because energy is lost from the bonds formed by the products.
- Endergonic is an anabolic reaction that consumes energy. It has a $+\Delta G$ because energy is required to break bonds.

The free energy (ΔG) gained or lost in a reaction can be calculated: $\Delta G = \Delta H - T\Delta S$.

Also, $\Delta G = \Delta G°' + 2.303RT \log([P]/[R])$ where:

- o R is the gas constant, 1.987 cal/mol
- o T is temperature in kelvins K = 273 + °C
- o $[P]$ is products
- o $[R]$ is the reactants.

Cotransport

In August 1960, Robert K. Crane presented for the first time his discovery of the sodium-glucose cotransport as the mechanism for intestinal glucose absorption. Crane's discovery of cotransport was the first ever proposal of flux coupling in biology and was the most important event concerning carbohydrate absorption in the 20th century.

Chemiosmotic Theory

One of the major triumphs of bioenergetics is Peter D. Mitchell's chemiosmotic theory of how protons in aqueous solution function in the production of ATP in cell organelles such as mitochondria. Other cellular sources of ATP such as glycolysis were understood first, but such processes for direct coupling of enzyme activity to ATP production are not the major source of useful chemical energy in most cells. Chemiosmotic coupling is the major energy producing process in most cells, being utilized in chloroplasts and several single celled organisms in addition to mitochondria.

Energy Balance

Energy balance is the biological homeostasis of energy in living systems. It is measured with the following equation:

Energy intake = internal heat produced + external work + storage.

Homeostasis

With regards to any given life system parameter, an organism may be a *conformer* or a *regulator*. On one hand, regulators try to maintain the parameter at a constant level over possibly wide ambient environmental variations. On the other hand, conformers allow the environment to determine the parameter. For instance, endothermic animals (mammals and birds) maintain a constant body temperature, while exothermic animals (almost all other organisms) exhibit wide body temperature variation.

Behavioural adaptations allow endothermic animals to exert some control over a given parameter. For instance, reptiles often rest on sun-heated rocks in the morning to raise their body temperature. Regulators are also responsive to external circumstances, however: if the same sun-baked boulder happens to host a ground squirrel, the animal's metabolism will adjust to the lesser need for internal heat production.

An advantage of homeostatic regulation is that it allows an organism to function effectively in a broad range of environmental conditions. For example, ectotherms tend to become sluggish at low temperatures, whereas a co-located endotherm may be fully active. That thermal stability comes at a price since an automatic regulation system requires additional energy. One reason snakes may eat only once a week is that they use much less energy to maintain homeostasis.

Most homeostatic regulation is controlled by the release of hormones into the bloodstream. However, other regulatory processes rely on simple diffusion to maintain a balance.

Homeostatic regulation extends far beyond the control of temperature. All animals also regulate their blood glucose, as well as the concentration of their blood. Mammals regulate their blood glucose with insulin and glucagon. The human body maintains glucose levels constant most of the day, even after a 24-hour fast. Even during long periods of fasting, glucose levels are reduced only very slightly. Insulin, secreted by the beta cells of the pancreas, effectively transports glucose to the body's cells by instructing those cells to keep more of the glucose for their own use. seeDynamic equilibrium.

If the glucose inside the cells is high the cells will convert it to the insoluble glycogen to prevent the soluble glucose interfering with cellular metabolism. Ultimately this lowers blood glucose levels, and Insulin helps to prevent hyperglycemia. When insulin is deficient or cells become resistant to it, diabetes occurs. Glucagon, secreted by the alpha cells of the pancreas, encourages cells to break down stored glycogen or convert non-carbohydrate carbon sources to glucose via gluconeogenesis, thus preventing hypoglycemia. The kidneys are used to remove excess water and ions from the blood. These are then expelled as urine. The kidneys perform a vital role in homeostatic regulation in mammals, removing excess water, salt, and urea from the blood. These are the body's main waste products.

Another homeostatic regulation occurs in the gut. Homeostasis of the gut is not fully understood but it is believed that Toll-like receptor (TLR) expression profiles contribute to it. Intestinal epithelial cells exhibit important factors that contribute to homeostasis:1) They have different cellular distribution of TLR's compared to the normal gut mucosa. An example of this is how TLR5 (activated by flagellin) can redistribute to the basolateral membrane, which is the perfect place where flagellin can be detected. 2) The enterocytes express high levels of TLR inhibitor Toll-interacting protein (TOLLIP).

TOLLIP is a human gene that is a part of the innate immune system and is highest in a healthy gut; it correlates to luminal bacterial load. 3) Surface enterocytes also express high levels of Interleukin-1 receptor (IL-1R)-containing inhibitory molecule. IL-1R are also referred to as single immunoglobulin IL-1R (SIGIRR).

Animals deficient in this are more susceptible to induced colitis, implying that SIGIRR might possibly play a role in tuning mucosal tolerance towards commensal flora. Nucleotide-binding oligomerisation domain containing 2 (NOD2) is suggested to have an effect on suppressing inflammatory cascades based on recent evidence. It is believed to modulate signals transmitted through TLRs, TLR3, 4, and 9 specifically. Mutation of it has resulted in Crohn's disease. Excessive T-helper 1 responses to resident flora in the gut are controlled by inhibiting the controlling influence of regulatory T cells and tolerance-inducing dendritic cells.

Sleep timing depends upon a balance between homeostatic sleep propensity, the need for sleep as a function of the amount of time elapsed since the last adequate sleep episode, and circadian rhythms that determine the ideal timing of a correctly structured and restorative sleep episode.

Control Mechanisms

All homeostatic control mechanisms have at least three interdependent components for the variable being regulated: The receptor is the sensing component that monitors and responds to changes in the environment. When the receptor senses a stimulus, it sends information to a control centre, the component that sets the range at which a variable is maintained. The control centre determines an appropriate response to the stimulus. In most homeostatic mechanisms the control centre is the brain. The control centre then sends signals to an effector, which can be muscles, organs or other structures that receive signals from the control centre. After receiving the signal, a change occurs to correct the deviation by either enhancing it with positive feedback or depressing it with negative feedback

Positive Feedback

Positive feedback mechanisms are designed to accelerate or enhance the output created by a stimulus that has already been activated.

Unlike negative feedback mechanisms that initiate to maintain or regulate physiological functions within a set and narrow range, the positive feedback mechanisms are designed to push levels out of normal ranges. To achieve this purpose, a series of events initiates a cascading process that builds to increase the effect of the stimulus. This process can be beneficial but is rarely used by the body due to risks of the acceleration's becoming uncontrollable.

One positive feedback example event in the body is blood platelet accumulation, which, in turn, causes blood clotting in response to a break or tear in the lining of blood vessels. Another example is the release of oxytocin to intensify the contractions that take place during childbirth.

Negative Feedback

Negative feedback mechanisms consist of reducing the output or activity of any organ or system back to its normal range of functioning. A good example of this is regulating blood pressure. Blood vessels can sense resistance of blood flow against the walls when blood pressure increases. The blood vessels act as the receptors and they relay this message to the brain. The brain then sends a message to the heart and blood vessels, both of which are the effectors. The heart rate would decrease as the blood vessels increase in diameter (or vasodilation). This change would cause the blood pressure to fall back to its normal range. The opposite would happen when blood pressure decreases, and would cause vasoconstriction.

Another important example is seen when the body is deprived of food. The body would then reset the metabolic set point to a lower than normal value. This would allow the body to continue to function, at a slower rate, even though the body is starving. Therefore, people who deprive themselves of food while trying to lose weight would find it easy to shed weight initially and much harder to lose more after. This is due to the body readjusting itself to a lower metabolic set point to allow the body to survive with its low supply of energy. Exercise can change this effect by increasing the metabolic demand.

Another good example of negative feedback mechanism is temperature control. The hypothalamus, which monitors the body temperature, is capable of determining even the slightest of variation of normal body temperature (37 degrees Celsius). Response to such variation could be stimulation of glands that produces sweat to reduce the temperature or signalling various muscles to shiver to increase body temperature.

Both feedbacks are equally important for the healthy functioning of one's body. Complications can arise if any of the two feedbacks are affected or altered in any way.

Homeostatic Imbalance

Many diseases are a result of disturbance of homeostasis, a condition known as homeostatic imbalance. As it ages, every organism will lose efficiency in its control systems. The inefficiencies gradually result in an unstable internal environment that increases the risk for illness. In addition, homeostatic imbalance is also responsible for the physical changes associated with aging. Even more serious than illness and other characteristics of aging is death. Heart failure has been seen where nominal negative feedback mechanisms become overwhelmed, and destructive positive feedback mechanisms then take over.

Diseases that result from a homeostatic imbalance include diabetes, dehydration, hypoglycemia, hyperglycemia, gout, and any disease caused by a toxin present in the bloodstream. All of these conditions result from the presence of an increased amount of a particular substance. In ideal circumstances, homeostatic control mechanisms should prevent this imbalance from occurring, but, in some people, the mechanisms do not work efficiently enough or the quantity of the substance exceeds the levels at which it can be managed. In these cases, medical intervention is necessary to restore the balance, or permanent damage to the organs may result.

Varieties

The Dynamic Energy Budget theory for metabolic organization delineates structure and (one or more) reserves in an organism. Its formulation is based on three forms of homeostasis:

- Strong homeostasis, whereas structure and reserve do not change in composition. Because the amount of reserve and structure can vary, this allows a particular change in the composition of the whole body (as explained by the Dynamic Energy Budget theory).

- Weak homeostasis, wherein the ratio of the amounts of reserve and structure becomes constant as long as food availability is constant, even when the organism grows. This means that the whole body composition is constant during growth in constant environments.
- Structural homeostasis, wherein the sub-individual structures grow in harmony with the whole individual; the relative proportions of the individuals remain constant.

Ecological

Historically, ecological succession was seen as having a stable end-stage called the climax, sometimes referred to as the 'potential biodiversity' of a site, shaped primarily by the local climate. This idea has been largely abandoned by modern ecologists in favour of nonequilibrium ideas of how ecosystems function, as most natural ecosystems experience disturbance at a rate that makes a "climax" community unattainable.

Only on small, isolated habitats known as ecological islands can the phenomenon be observed. One such case study is the island of Krakatoa after its major eruption in 1883: the established stable homeostasis of the previous forest climax ecosystem was destroyed, and all life was eliminated from the island. In the years after the eruption, Krakatoa went through a sequence of ecological changes in which successive groups of new plant or animal species followed one another, leading to increasing biodiversity and eventually culminating in a re-established climax community.

This *ecological succession* on Krakatoa occurred in a number of stages; a *sere* is defined as "a stage in a sequence of events by which succession occurs". The complete chain of seres leading to a climax is called a *prisere.*

In the case of Krakatoa, the island reached its climax community, with eight hundred different recorded species, in 1983, one hundred years after the eruption that cleared all life off the island. Evidence confirms that this number has been homeostatic for some time, with the introduction of new species rapidly leading to elimination of old ones. The evidence of Krakatoa, and other disturbed island ecosystems, has confirmed many principles of Island Biogeography, mimicking general principles of ecological succession albeit in a virtually closed system comprised almost exclusively of endemic species.

Biosphere

In the Gaia hypothesis, James Lovelock stated that the entire mass of living matter on Earth (or any planet with life) functions as a vast homeostatic superorganism that actively modifies its planetary environment to produce the environmental conditions necessary for its own survival.

Whether this sort of system is present on Earth is still open to debate. However, some relatively simple homeostatic mechanisms are generally accepted. For example, when atmospheric carbon dioxide levels rise, certain plants are able to grow better and thus act to remove more carbon dioxide from the atmosphere. When sunlight is plentiful and atmospheric temperature climbs, the phytoplankton of the ocean surface waters thrive and produce more dimethyl sulfide, DMS. The DMS molecules act as cloud condensation nuclei, which produce more clouds, and thus increase the atmospheric albedo, and this feeds back to lower the temperature of the atmosphere.

As scientists discover more about Earth, vast numbers of positive and negative feedback loops are being discovered, that, together, maintain a metastable condition, sometimes within very broad range of environmental conditions. Environmental pressure, such as competition or change in temperature, can lead to adaptation/extinction of species over time.

Reactive

Example of use: "Reactive homeostasis is an immediate homeostasic response to a challenge such as predation."

However, *any* homeostasis is impossible without reaction-because homeostasis is and must be a "feedback" phenomenon.

The phrase "reactive homeostasis" is simply short for "reactive compensation reestablishing homeostasis", that is to say, "reestablishing a point of homeostasis."-it should not be confused with a separate *kind* of homeostasis or a distinct phenomenon *from* homeostasis; it is simply the compensation (or compensatory) phase of homeostasis.

Glycolysis and Fermentation

Glycolysis (from *glycose*, an older term for glucose + *-lysis* degradation) is the metabolic pathway that converts glucose $C_6H_{12}O_6$, into pyruvate, $CH_3COCOO^- + H^+$. The free energy released in this process is used to form the high-energy compounds ATP (adenosine triphosphate) and NADH (reduced nicotinamide adenine dinucleotide).

Glycolysis is a definite sequence of ten reactions involving ten intermediate compounds (one of the steps involves two intermediates). The intermediates provide entry points to glycolysis. For example, most monosaccharides, such as fructose, glucose, and galactose, can be converted to one of these intermediates. The intermediates may also be directly useful. For example, the intermediate dihydroxyacetone phosphate is a source of the glycerol that combines with fatty acids to form fat. Glycolysis is thought to be the archetype of a universal metabolic pathway. It occurs, with variations, in nearly all organisms, both aerobic and anaerobic. The wide occurrence of glycolysis indicates that it is one of the most ancient known metabolic pathways.

The most common type of glycolysis is the *Embden-Meyerhof-Parnas pathway (EMP pathway)*, which was first discovered by Gustav Embden, Otto Meyerhof and Jakub Karol Parnas. Glycolysis also refers to other pathways, such as the *Entner–Doudoroff pathway* and various heterofermentative and homofermentative pathways. Dietary carbohydrate from which humans gain energy enter the body in complex forms, such as disaccharides and the polymers starch (amylose and amylopectin) and glycogen. The polymer cellulose is also consumed but not digested. The first step in the metabolism of digestible carbohydrate is the conversion of the higher polymers to simpler, soluble forms that can be transported across the intestinal wall and delivered to the tissues.

The breakdown of polymeric sugars begins in the mouth. Saliva has a slightly acidic pH of 6.8 and contains lingual amylase that begins the digestion of carbohydrates. The action of lingual amylase is limited to the area of the mouth and the esophagus; it is virtually inactivated by the much stronger acid pH of the stomach. Once the food has arrived in the stomach, acid hydrolysis contributes to its degradation; specific gastric proteases and lipases aid this process for proteins and fats, respectively. The mixture of gastric secretions, saliva, and food, known collectively as chyme, moves to the small intestine.

The main polymeric-carbohydrate digesting enzyme of the small intestine is α-amylase. This enzyme is secreted by the pancreas and has the same activity as salivary amylase, producing disaccharides and trisaccharides. The latter are converted to monosaccharides by intestinal saccharidases, including maltases that hydrolyze di- and trisaccharides, and the more specific disaccharidases, sucrase, lactase, and trehalase.

The net result is the almost complete conversion of digestible carbohydrate to its constituent monosaccharides. The resultant glucose and other simple carbohydrates are transported across the intestinal wall to the hepatic portal vein and then to liver parenchymal cells and other tissues. There they are converted to fatty acids, amino acids, and glycogen, or else oxidized by the various catabolic pathways of cells.

Oxidation of glucose is known as glycolysis. Glucose is oxidized to either lactate or pyruvate. Under aerobic conditions, the dominant product in most tissues is pyruvate and the pathway is known as aerobic glycolysis. When oxygen is depleted, as for instance during prolonged vigorous exercise, the dominant glycolytic product in many tissues is lactate and the process is known as anaerobic glycolysis.

The Energy Derived from Glucose Oxidation

Aerobic glycolysis of glucose to pyruvate, requires two equivalents of ATP to activate the process, with the subsequent production of four equivalents of ATP and two equivalents of NADH. Thus, conversion of one mole of glucose to two moles of pyruvate is accompanied by the net production of two moles each of ATP and NADH.

$$\text{Glucose} + 2\ \text{ADP} + 2\ \text{NAD}^+ + 2\ \text{P}_i \rightarrow$$
$$2\ \text{Pyruvate} + 2\ \text{ATP} + 2\ \text{NADH} + 2\ \text{H}^+$$

The NADH generated during glycolysis is used to fuel mitochondrial ATP synthesis via oxidative phosphorylation, producing either two or three equivalents of ATP depending upon whether the glycerol phosphate shuttle or the malate-aspartate shuttle is used to transport the electrons from cytoplasmic NADH into the mitochondria.

The malate-aspartate shuttle is the principal mechanism for the movement of reducing equivalents (in the form of NADH) from the cytoplasm to the mitochondria. The glycolytic pathway is a primary source of NADH. Within the mitochodria the electrons of NADH can be coupled to ATP production during the process of oxidative phosphorylation.

The electrons are "carried" into the mitochondria in the form of malate. Cytoplasmic malate dehydrogenase (MDH) reduces oxaloacetate (OAA) to malate while oxidizing NADH to NAD$^+$. Malate then enters the mitochondria where the reverse reaction is carried out by mitochondrial MDH. Movement of mitochondrial OAA to the cytoplasm to maintain this cycle requires it be transaminated to aspartate (Asp, D) with the amino group being donated by glutamate (Glu, E).

The Asp then leaves the mitochondria and enters the cytoplasm. The deamination of glutamate generates α-ketoglutarate (α-KG) which leaves the mitochondria for the cytoplasm.

All the participants in the cycle are present in the proper cellular compartment for the shuttle to function duc to concentration dependent movement. When the energy level of the cell rises the rate of mitochondrial oxidation of NADH to NAD$^+$ declines and therefore, the shuttle slows. G3PDH is glyceraldehyde-3-phosphate dehydrogenase.

The glycerol phosphate shuttle is a secondary mechanism for the transport of electrons from cytosolic NADH to mitochondrial carriers of the oxidative phosphorylation pathway. The primary cytoplasmic NADH electron shuttle is the malate-aspartate shuttle.

Two enzymes are involved in this shuttle. One is the cytosolic version of the enzyme glycerol-3-phosphate dehydrogenase (glycerol-3-PDH) which has as one substrate, NADH. The second is the mitochondrial form of the enzyme which has as one of its' substrates, FAD^+. The net result is that there is a continual conversion of the glycolytic intermediate, DHAP and glycerol-3-phosphate with the concomitant transfer of the electrons from reduced cytosolic NADH to mitochondrial oxidized FAD^+. Since the electrons from mitochondrial $FADH_2$ feed into the oxidative phosphorylation pathway at coenzyme Q (as opposed to NADH-ubiquinone oxidoreductase [complex I]) only 2 moles of ATP will be generated from glycolysis. G3PDH is glyceraldehyde-3-phoshate dehydrogenase.

The net yield from the oxidation of 1 mole of glucose to 2 moles of pyruvate is, therefore, either 6 or 8 moles of ATP. Complete oxidation of the 2 moles of pyruvate, through the TCA cycle, yields an additional 30 moles of ATP; the total yield, therefore being either 36 or 38 moles of ATP from the complete oxidation of 1 mole of glucose to CO_2 and H_2O.

The Individual Reactions of Glycolysis

The pathway of glycolysis can be seen as consisting of 2 separate phases. The first is the chemical priming phase requiring energy in the form of ATP, and the second is considered the energy-yielding phase. In the first phase, 2 equivalents of ATP are used to convert glucose to fructose 1,6-bisphosphate (F1,6BP). In the second phase F1,6BP is degraded to pyruvate, with the production of 4 equivalents of ATP and 2 equivalents of NADH.

Pathway of glycolysis from glucose to pyruvate. Substrates and products are in blue, enzymes are in green. The two high energy intermediates whose oxidations are coupled to ATP synthesis are shown in red (1,3-bisphosphoglycerate and phosphoenolpyruvate).

The Hexokinase Reaction

The ATP-dependent phosphorylation of glucose to form glucose 6-phosphate (G6P) is the first reaction of glycolysis, and is catalysed by tissue-specific isoenzymes known as hexokinases. The phosphorylation accomplishes two goals: First, the hexokinase reaction converts nonionic glucose into an anion that is trapped in the cell, since cells lack transport systems for phosphorylated sugars. Second, the otherwise biologically inert glucose becomes activated into a labile form capable of being further metabolized.

Four mammalian isozymes of hexokinase are known (Types I–IV), with the Type IV isozyme often referred to as glucokinase. Glucokinase is the form of the enzyme found in hepatocytes and pancreatic β-cells. The high K_m of glucokinase for glucose means that this enzyme is saturated only at very high concentrations of substrate.

This feature of hepatic glucokinase allows the liver to buffer blood glucose. After meals, when postprandial blood glucose levels are high, liver glucokinase is significantly active, which causes the liver preferentially to trap and to store circulating glucose. When blood glucose falls to very low levels, tissues such as liver and kidney, which contain glucokinases but are not highly dependent on glucose, do not continue to use the meager glucose supplies that remain available. At the same time, tissues such as the brain, which are critically dependent on glucose, continue to scavenge blood glucose using their low K_m hexokinases, and as a consequence their viability is protected. Under various conditions of glucose deficiency, such as long periods between meals, the liver is stimulated to supply the blood with glucose through the pathway of gluconeogenesis.

The levels of glucose produced during gluconeogenesis are insufficient to activate glucokinase, allowing the glucose to pass out of hepatocytes and into the blood.

The regulation of hexokinase and glucokinase activities is also different. Hexokinases I, II, and III are allosterically inhibited by product (G6P) accumulation, whereas glucokinases are not. The latter further insures liver accumulation of glucose stores during times of glucose excess, while favouring peripheral glucose utilization when glucose is required to supply energy to peripheral tissues.

Phosphohexose Isomerase

The second reaction of glycolysis is an isomerization, in which G6P is converted to fructose 6-phosphate (F6P). The enzyme catalysing this reaction is phosphohexose isomerase. The reaction is freely reversible at normal cellular concentrations of the two hexose phosphates and thus catalyzes this interconversion during glycolytic carbon flow and during gluconeogenesis.

6-Phosphofructo-1-Kinase (Phosphofructokinase-1, PFK-1):

The next reaction of glycolysis involves the utilization of a second ATP to convert F6P to fructose 1,6-bisphosphate (F1,6BP). This reaction is catalysed by 6-phosphofructo-1-kinase, better known as phosphofructokinase-1 or PFK-1. This reaction is not readily reversible because of its large positive free energy ($\Delta G^{0'} = +5.4$ kcal/mol) in the reverse direction. Nevertheless, fructose units readily flow in the reverse (gluconeogenic) direction because of the ubiquitous presence of the hydrolytic enzyme, fructose-1,6-bisphosphatase (F-1,6-BPase).

The presence of these two enzymes in the same cell compartment provides an example of a metabolic futile cycle, which if unregulated would rapidly deplete cell energy stores. However, the activity of these two enzymes is so highly regulated that PFK-1 is considered to be the rate-limiting enzyme of glycolysis and F-1,6-BPase is considered to be the rate-limiting enzyme in gluconeogenesis.

Aldolase

Aldolase catalyses the hydrolysis of F1,6BP into two 3-carbon products: dihydroxyacetone phosphate (DHAP) and glyceraldehyde 3-phosphate (G3P). The aldolase reaction proceeds readily in the reverse direction, being utilized for both glycolysis and gluconeogenesis.

Triose Phosphate Isomerase

The two products of the aldolase reaction equilibrate readily in a reaction catalysed by triose phosphate isomerase. Succeeding reactions of glycolysis utilize G3P as a substrate; thus, the aldolase reaction is pulled in the glycolytic direction by mass action principals.

Glyceraldehyde-3-Phosphate Dehydrogenase

The second phase of glucose catabolism features the energy-yielding glycolytic reactions that produce ATP and NADH. In the first of these reactions, glyceraldehyde-3-P dehydrogenase (G3PDH) catalyzes the NAD^+-dependent oxidation of G3P to 1,3-bisphosphoglycerate (1,3BPG) and NADH. The G3PDH reaction is reversible, and the same enzyme catalyzes the reverse reaction during gluconeogenesis.

Phosphoglycerate Kinase

The high-energy phosphate of 1,3-BPG is used to form ATP and 3-phosphoglycerate (3PG) by the enzyme phosphoglycerate kinase. Note that this is the only reaction of glycolysis or gluconeogenesis that involves ATP and yet is reversible under normal cell conditions. Associated with the phosphoglycerate kinase pathway is an important reaction of erythrocytes, the formation of 2,3-bisphosphoglycerate, 2,3BPG by the enzyme bisphosphoglycerate mutase. 2,3BPG is an important regulator of hemoglobin's affinity for oxygen. Note that 2,3-bisphosphoglycerate phosphatase degrades 2,3BPG to 3-phosphoglycerate, a normal intermediate of glycolysis. The 2,3BPG shunt thus operates with the expenditure of 1 equivalent of ATP per triose passed through the shunt. The process is not reversible under physiological conditions. The pathway for 2,3-bisphosphoglycerate (2,3-BPG) synthesis within erythrocytes. Synthesis of 2,3-BPG represents a major reaction pathway for the consumption of glucose in erythrocytes. The synthesis of 2,3-BPG in erythrocytes is critical for controlling hemoglobin affinity for oxygen. Note that when glucose is oxidized by this pathway the erythrocyte loses the ability to gain 2 moles of ATP from glycolytic oxidation of 1,3-BPG to 3-phosphoglycerate via the phosphoglycerate kinase reaction.

Phosphoglycerate Mutase and Enolase

The remaining reactions of glycolysis are aimed at converting the relatively low energy phosphoacyl-ester of 3PG to a high-energy form and harvesting the phosphate as ATP. The 3PG is first converted to 2PG by phosphoglycerate mutase and the 2PG conversion to phosphoenoylpyruvate (PEP) is catalysed by enolase.

Pyruvate Kinase

The final reaction of aerobic glycolysis is catalysed by the highly regulated enzyme pyruvate kinase (PK). In this strongly exergonic reaction, the high-energy phosphate of PEP is conserved as ATP. The loss of phosphate by PEP leads to the production of pyruvate in an unstable enol form, which spontaneously tautomerizes to the more stable, keto form of pyruvate. This reaction contributes a large proportion of the free energy of hydrolysis of PEP.

There are two distinct genes encoding PK activity. One is located on chromosome 1 and encodes the liver and erythrocyte PK proteins (identified as the PKLR gene) and the other is located on chromosome 15 and encodes the muscle PK proteins (identified as the PKM gene). The muscle PKM gene directs the synthesis of two isoforms of muscle PK termed PK-M1 and PK-M2. Deficiencies in the PKLR gene are the cause of the most common form of inherited non-spherocytic anemia.

Anaerobic Glycolysis

Under aerobic conditions, pyruvate in most cells is further metabolized via the TCA cycle. Under anaerobic conditions and in erythrocytes under aerobic conditions, pyruvate is converted to lactate by the enzyme lactate dehydrogenase (LDH), and the lactate is transported out of the cell into the circulation. The conversion of pyruvate to lactate, under anaerobic conditions, provides the cell with a mechanism for the oxidation of NADH (produced during the G3PDH reaction) to NAD^+ which occurs during the LDH catalysed reaction. This reduction is required since NAD^+ is a necessary substrate for G3PDH, without which glycolysis will cease. Normally, during aerobic glycolysis the electrons of cytoplasmic NADH are transferred to mitochondrial carriers of the oxidative phosphorylation pathway generating a continuous pool of cytoplasmic NAD^+.

Aerobic glycolysis generates substantially more ATP per mole of glucose oxidized than does anaerobic glycolysis. The utility of anaerobic glycolysis, to a muscle cell when it needs large amounts of energy, stems from the fact that the rate of ATP production from glycolysis is approximately 100X faster than from oxidative phosphorylation. During exertion muscle cells do not need to energize anabolic reaction pathways. The requirement is to generate the maximum amount of ATP, for muscle contraction, in the shortest time frame. This is why muscle cells derive almost all of the ATP consumed during exertion from anaerobic glycolysis.

Regulation of Glycolysis

The reactions catalysed by hexokinase, PFK-1 and PK all proceed with a relatively large free energy decrease. These non-equilibrium reactions of glycolysis would be ideal candidates for regulation of the flux through glycolysis. Indeed, *in vitro* studies have shown all three enzymes to be allosterically controlled.

Regulation of hexokinase, however, is not the major control point in glycolysis. This is due to the fact that large amounts of G6P are derived from the breakdown of glycogen (the predominant mechanism of carbohydrate entry into glycolysis in skeletal muscle) and, therefore, the hexokinase reaction is not necessary. Regulation of PK is important for reversing glycolysis when ATP is high in order to activate gluconeogenesis. As such this enzyme catalysed reaction is not a major control point in glycolysis. The rate limiting step in glycolysis is the reaction catalysed by PFK-1.

PFK-1 is a tetrameric enzyme that exist in two conformational states termed R and T that are in equilibrium. ATP is both a substrate and an allosteric inhibitor of PFK-1. Each subunit has two ATP binding sites, a substrate site and an inhibitor site. The substrate site binds ATP equally well when the tetramer is in either conformation.

The inhibitor site binds ATP essentially only when the enzyme is in the T state. F6P is the other substrate for PFK-1 and it also binds preferentially to the R state enzyme. At high concentrations of ATP, the inhibitor site becomes occupied and shifting the equilibrium of PFK-1 conformation to that of the T state decreasing PFK-1's ability to bind F6P. The inhibition of PFK-1 by ATP is overcome by AMP which binds to the R state of the enzyme and, therefore, stabilizes the conformation of the enzyme capable of binding F6P. The most important allosteric regulator of both glycolysis and gluconeogenesis is fructose 2,6-bisphosphate, F2,6BP, which is not an intermediate in glycolysis or in gluconeogenesis.

Regulation of glycolysis and gluconeogenesis by fructose 2,6-bisphosphate (F2,6BP). The major sites for regulation of glycolysis and gluconeogenesis are the phosphofructokinase-1 (PFK-1) and fructose-1,6-bisphosphatase (F-1,6-BPase) catalysed reactions. PFK-2 is the kinase activity and F-2,6-BPase is the phosphatase activity of the bi-functional regulatory enzyme, phosphofructokinase-2/fructose-2,6-bisphosphatase. PKA is cAMP-dependent protein kinase which phosphorylates PFK-2/F-

2,6-BPase turning on the phosphatase activity. (+ve) and (-ve) refer to positive and negative activities, respectively.

The synthesis of F2,6BP is catalysed by the bifunctional enzyme phosphofructokinase-2/fructose-2,6-bisphosphatase (PFK-2/F-2,6-BPase). In the nonphosphorylated form the enzyme is known as PFK-2 and serves to catalyse the synthesis of F2,6BP by phosphorylating fructose 6-phosphate. The result is that the activity of PFK-1 is greatly stimulated and the activity of F-1,6-BPase is greatly inhibited.

Under conditions where PFK-2 is active, fructose flow through the PFK-1/F-1,6-BPase reactions takes place in the glycolytic direction, with a net production of F1,6BP. When the bifunctional enzyme is phosphorylated it no longer exhibits kinase activity, but a new active site hydrolyzes F2,6BP to F6P and inorganic phosphate. The metabolic result of the phosphorylation of the bifunctional enzyme is that allosteric stimulation of PFK-1 ceases, allosteric inhibition of F-1,6-BPase is eliminated, and net flow of fructose through these two enzymes is gluconeogenic, producing F6P and eventually glucose. The interconversion of the bifunctional enzyme is catalysed by cAMP-dependent protein kinase (PKA), which in turn is regulated by circulating peptide hormones.

When blood glucose levels drop, pancreatic insulin production falls, glucagon secretion is stimulated, and circulating glucagon is highly increased. Hormones such as glucagon bind to plasma membrane receptors on liver cells, activating membrane-localized adenylate cyclase leading to an increase in the conversion of ATP to cAMP. cAMP binds to the regulatory subunits of PKA, leading to release and activation of the catalytic subunits. PKA phosphorylates numerous enzymes, including the bifunctional PFK-2/F-2,6-BPase. Under these conditions the liver stops consuming glucose and becomes metabolically gluconeogenic, producing glucose to reestablish normoglycemia.

Representative pathway for the activation of cAMP-dependent protein kinase (PKA). In this example glucagon binds to its' cell-surface receptor, thereby activating the receptor. Activation of the receptor is coupled to the activation of a receptor-coupled G-protein (GTP-binding and hydrolyzing protein) composed of 3 subunits. Upon activation the alpha subunit dissociates and binds to and activates adenylate cyclase. Adenylate cylcase then converts ATP to cyclic-AMP (cAMP). The cAMP thus produced then binds to the regulatory subunits of PKA leading to dissociation of the associated catalytic subunits. The catalytic subunits are inactive until dissociated from the regulatory subunits. Once released the catalytic subunits of PKA phosphorylate numerous substrate using ATP as the phosphate donor.

Regulation of glycolysis also occurs at the step catalysed by pyruvate kinase, (PK). The liver enzyme has been most studied *in vitro*. This enzyme is inhibited by ATP and acetyl-CoA and is activated by F1,6BP. The inhibition of PK by ATP is similar to the effect of ATP on PFK-1. The binding of ATP to the inhibitor site reduces its affinity for PEP. The liver enzyme is also controlled at the level of synthesis. Increased carbohydrate ingestion induces the synthesis of PK resulting in elevated cellular levels of the enzyme.

A number of PK isozymes have been described. The liver isozyme (L-type), characteristic of a gluconeogenic tissue, is regulated via phosphorylation by PKA, whereas the M-type isozyme found in brain, muscle, and other glucose requiring tissue is unaffected by PKA. As a consequence of these differences, blood glucose levels and associated hormones can regulate the balance of liver gluconeogenesis and glycolysis while muscle metabolism remains unaffected.

In erythrocytes, the fetal PK isozyme has much greater activity than the adult isozyme; as a result, fetal erythrocytes have comparatively low concentrations of glycolytic intermediates. Because of the low steady-state concentration of fetal 1,3BPG, the 2,3BPG shunt is greatly reduced in fetal cells and little 2,3BPG is formed. Since 2,3BPG is a negative effector of hemoglobin affinity for oxygen, fetal erythrocytes have a higher oxygen affinity than maternal erythrocytes. Therefore, transfer of oxygen from maternal hemoglobin to fetal hemoglobin is favoured, assuring the fetal oxygen supply. In the newborn, an erythrocyte isozyme of the M-type with comparatively low PK activity displaces the fetal type, resulting in an accumulation of glycolytic intermediates. The increased 1,3BPG levels activate the 2,3BPG shunt, producing 2,3BPG needed to regulate oxygen binding to hemoglobin.

Genetic diseases of adult erythrocyte PK are known in which the kinase is virtually inactive. The erythrocytes of affected individuals have a greatly reduced capacity to make ATP and thus do not have sufficient ATP to perform activities such as ion pumping and maintaining osmotic balance. These erythrocytes have a short half-life, lyse readily, and are responsible for some cases of hereditary hemolytic anemia.

The liver PK isozyme is regulated by phosphorylation, allosteric effectors, and modulation of gene expression. The major allosteric effectors are F1,6BP, which stimulates PK activity by decreasing its K_m for PEP, and for the negative effector, ATP. Expression of the liver PK gene is strongly influenced by the quantity of carbohydrate in the diet, with high-carbohydrate diets inducing up to 10-fold increases in PK concentration as compared to low carbohydrate diets. Liver PK is phosphorylated and inhibited by PKA, and thus it is under hormonal control similar to that described earlier for PFK-2.

Muscle PK (M-type) is not regulated by the same mechanisms as the liver enzyme. Extracellular conditions that lead to the phosphorylation and inhibition of liver PK, such as low blood glucose and high levels of circulating glucagon, do not inhibit the muscle enzyme. The result of this differential regulation is that hormones such as glucagon and epinephrine favour liver gluconeogenesis by inhibiting liver glycolysis, while at the same time, muscle glycolysis can proceed in accord with needs directed by intracellular conditions.

Metabolic Fates of Pyruvate

Pyruvate is the branch point molecule of glycolysis. The ultimate fate of pyruvate depends on the oxidation state of the cell. In the reaction catalysed by G3PDH a molecule of NAD^+ is reduced to NADH. In order to maintain the re-dox state of the cell, this NADH must be re-oxidized to NAD^+. During aerobic glycolysis this occurs in the mitochondrial electron transport chain generating ATP. Thus, during aerobic glycolysis ATP is generated from oxidation of glucose directly at the PGK and PK reactions as well as indirectly by re-oxidation of NADH in the oxidative phosphorylation pathway. Additional NADH molecules are generated during the complete aerobic oxidation of pyruvate in the TCA cycle. Pyruvate enters the TCA cycle in the form of acetyl-CoA which is the product of the pyruvate dehydrogenase reaction. The fate of pyruvate during anaerobic glycolysis is reduction to lactate.

Lactate Metabolism

During anaerobic glycolysis, that period of time when glycolysis is proceeding at a high rate (or in anaerobic organisms), the oxidation of NADH occurs through the reduction of an organic substrate. Erythrocytes and skeletal muscle (under conditions of exertion) derive all of their ATP needs through anaerobic glycolysis. The large quantity of NADH produced is oxidized by reducing pyruvate to lactate. This reaction is carried out by lactate dehydrogenase, (LDH). The lactate produced during anaerobic glycolysis diffuses from the tissues and is transported to highly aerobic tissues such as cardiac muscle and liver. The lactate is then oxidized to pyruvate in these cells by LDH and the pyruvate is further oxidized in the TCA cycle. If the energy level in these cells is high the carbons of pyruvate will be diverted back to glucose via the gluconeogenesis pathway.

Mammalian cells contain two distinct types of LDH subunits, termed M and H. Combinations of these different subunits generates LDH isozymes with different characteristics. The H type subunit predominates in aerobic tissues such as heart muscle (as the H4 tetramer) while the M subunit predominates in anaerobic tissues such as skeletal muscle as the M4 tetramer). H4 LDH has a low K_m for pyruvate and also is inhibited by high levels of pyruvate. The M4 LDH enzyme has a high K_m for pyruvate and is not inhibited by pyruvate. This suggests that the H-type LDH is utilized for oxidizing lactate to pyruvate and the M-type the reverse.

Ethanol Metabolism

Metabolism of ethanol occurs primarily in the liver and involves at least three enzymatic pathways. The most significant pathway, responsible for the bulk of ethanol metabolism, is that initiated by alcohol dehydrogenase, ADH. The second major pathway for ethanol metabolism is the microsomal ethanol oxidizing system (MEOS) which involves the cytochrome P450 enzyme CYP2E1 and requires NADPH instead of NAD^+ as for ADH. The MEOS pathway is induced in individuals who chronically consume alcohol. The third pathway involves a non-oxidative pathway catalysed by fatty acid ethyl ester (FAEE) synthase. This latter pathway results in the formation of fatty acid ethyl esters and takes place primarily in the liver and pancreas, both of which are highly susceptible to the toxic effects of alcohol. Animal cells (primarily hepatocytes) contain cytosolic ADH which oxidizes ethanol to acetaldehyde. Acetaldehyde then enters the mitochondria where it is oxidized to acetate by acetaldehyde dehydrogenase (AcDH).

In humans there are multiple isoforms of ADH encoded for by at least seven different ADH genes. Functional ADH exists as a dimer and can be classified by five distinct classes (denoted I–V). The hepatic form of ADH is class I and consists of A, B, and C subunits encoded by three different ADH genes (ADH1A, ADH1B, and ADH1C). It should be pointed out that humans evolved to express multiple ADH genes and isoforms, not for metabolism of ethanol, but to metabolize naturally occurring alcohols found in foods as well as those produced by intestinal bacteria. As an example, one form of ADH is responsible for the metabolism of retinol (vitamin A) to retinaldehyde which is the form of vitamin A necessary for vision.

Acetaldehyde forms adducts with proteins, nucleic acids and other compounds, the results of which are the toxic side effects (the hangover) that are associated with alcohol consumption. The ADH and AcDH catalysed reactions also leads to the reduction of NAD^+ to NADH. The metabolic effects of ethanol intoxication stem from the actions of ADH and AcDH and the resultant cellular imbalance in the NADH/NAD^+. The NADH produced in the cytosol by ADH must be reduced back to NAD^+ via either the malate-aspartate shuttle or the glycerol-phosphate shuttle. Thus, the ability of an individual to metabolize ethanol is dependent upon the capacity of hepatocytes to carry out either of these 2 shuttles, which in turn is affected by the rate of the TCA cycle in the mitochondria whose rate of function is being impacted by the NADH produced by the AcDH reaction.

The reduction in NAD^+ impairs the flux of glucose through glycolysis at the glyceraldehyde-3-phosphate dehydrogenase reaction, thereby limiting energy production. Additionally, there is an increased rate of hepatic lactate production due to the effect of increased NADH on direction of the hepatic lactate dehydrogenase (LDH) reaction. This reversal of the LDH reaction in hepatocytes diverts pyruvate from gluconeogenesis leading to a reduction in the capacity of the liver to deliver glucose to the blood.

In addition to the negative effects of the altered $NADH/NAD^+$ ratio on hepatic gluconeogenesis, fatty acid oxidation is also reduced as this process requires NAD^+ as a cofactor. In fact the opposite is true, fatty acid synthesis is increased and there is an increase in triacylglyceride production by the liver. In the mitochondria, the production of acetate from acetaldehyde leads to increased levels of acetyl-CoA. Since the increased generation of NADH also reduces the activity of the TCA cycle, the acetyl-CoA is diverted to fatty acid synthesis. The reduction in cytosolic NAD^+ leads to reduced activity of glycerol-3-phosphate dehydrogenase (in the glycerol 3-phosphate to DHAP direction) resulting in increased levels of glycerol 3-phosphate which is the backbone for the synthesis of the triacylglycerides. Both of these two events lead to fatty acid deposition in the liver leading to fatty liver syndrome.

The MEOS pathway for ethanol metabolism is induced in chronic alcoholics and may contribute to the tolerance to ethanol observed in alcoholics. The activity of CYP2E1 is also essential in the metabolism of several xenobiotics. Therefore, the increased level of expression of this enzyme in alcoholics can have a significant impact on the production of toxic metabolites and this is thought to contribute to ethanol-induced liver injury. Of particular clinical significance is that increased levels of CYP2E1 result in accelerated conversion of the non-steroidal anti-inflammatory (NSAID) acetaminophen (Tylenol®) into toxic metabolites.

Metabolism of ethanol by CYP2E1 also results in a significant increase in free radical and acetaldehyde production which, in turn, diminish reduced glutathione (GSH) and other defence systems against oxidative stress leading to further hepatocyte damage. Along with increased CYP2E1 activity there is an induction of microsomal enzymes involved in lipoprotein production, resulting in hyperlipemia which contributes to the fatty liver syndrome.

Regulation of Blood Glucose Levels

If for no other reason, it is because of the demands of the brain for oxidizable glucose that the human body exquisitely regulates the level of glucose circulating in the blood. This level is maintained in the range of 5mM.

Nearly all carbohydrates ingested in the diet are converted to glucose following transport to the liver. Catabolism of dietary or cellular proteins generates carbon atoms that can be utilized for glucose synthesis via gluconeogenesis. Additionally, other tissues besides the liver that incompletely oxidize glucose (predominantly skeletal muscle and erythrocytes) provide lactate that can be converted to glucose via gluconeogenesis.

Maintenance of blood glucose homeostasis is of paramount importance to the survival of the human organism. The predominant tissue responding to signals that indicate reduced or elevated blood glucose levels is the liver. Indeed, one of the most important functions of the liver is to produce glucose for the circulation. Both elevated and reduced levels of blood glucose trigger hormonal responses to initiate pathways designed to restore glucose homeostasis.

Low blood glucose triggers release of glucagon from pancreatic α-cells. High blood glucose triggers release of insulin from pancreatic β-cells. Additional signals, ACTH and growth hormone, released from the pituitary act to increase blood glucose by inhibiting uptake by extrahepatic tissues. Glucocorticoids also act to increase blood glucose levels by inhibiting glucose uptake. Cortisol, the major glucocorticoid released from the adrenal cortex, is secreted in response to the increase in circulating ACTH. The adrenal medullary hormone, epinephrine, stimulates production of glucose by activating glycogenolysis in response to stressful stimuli.

Glucagon binding to its' receptors on the surface of liver cells triggers an increase in cAMP production leading to an increased rate of glycogenolysis by activating glycogen phosphorylase via the PKA-mediated cascade. This is the same response hepatocytes have to epinephrine release. The resultant increased levels of G6P in hepatocytes is hydrolysed to free glucose, by glucose-6-phosphatase, which then diffuses to the blood. The glucose enters extrahepatic cells where it is re-phosphorylated by hexokinase.

Since muscle and brain cells lack glucose-6-phosphatase, the glucose-6-phosphate product of hexokinase is retained and oxidized by these tissues. In opposition to the cellular responses to glucagon (and epinephrine on hepatocytes), insulin stimulates extrahepatic uptake of glucose from the blood and inhibits glycogenolysis in extrahepatic cells and conversely stimulates glycogen synthesis. As the glucose enters hepatocytes it binds to and inhibits glycogen phosphorylase activity. The binding of free glucose stimulates the de-phosphorylation of phosphorylase thereby, inactivating it. Why is it that the glucose that enters hepatocytes is not immediately phosphorylated and oxidized? Liver cells contain an isoform of hexokinase called glucokinase. Glucokinase has a much lower affinity for glucose than does hexokinase. Therefore, it is not fully active at the physiological ranges of blood glucose. Additionally, glucokinase is not inhibited by its product G6P, whereas, hexokinase is inhibited by G6P.

Hepatocytes, unlike most other cells, are freely permeable to glucose and are, therefore, essentially unaffected by the action of insulin at the level of increased glucose uptake. When blood glucose levels are low, the liver does not compete with other tissues for glucose since the extrahepatic uptake of glucose is stimulated in response to insulin. Conversely, when blood glucose levels are high extrahepatic needs are satisfied and the liver takes up glucose for conversion into glycogen for future needs. Under conditions of high blood glucose, liver glucose levels will be high and the activity of glucokinase will be elevated. The G6P produced by glucokinase is rapidly converted to G1P by phosphoglucomutase, where it can then be incorporated into glycogen.

Role of the Kidney in Blood Glucose Control

Although the liver is the major site of glucose homeostasis, the kidney plays a vital role in the overall process of regulating the level of blood glucose. The kidney carries out gluconeogenesis primarily using the carbon skeleton of glutamine and while so doing allows for the elimination of waste nitrogen and maintaining plasma pH balance.

In addition to carrying out gluconeogenesis, the kidney regulates blood glucose levels via its ability to excrete glucose via glomerular filtration as well as to reabsorb the filtered glucose in the proximal convoluted tubules. The average adult will kidneys will filter around 180gm of glucose per day. Of this amount less than 1% is excreted in the urine due to efficient reabsorption. This reabsorption process is critical for maintaining blood glucose homeostasis and for retaining important calories for energy production.

Transport of glucose from the tubule into the tubular epithelial cells is carried out by specialized transport proteins termed sodium-glucose co-transporters (SGLTs). The SGLTs represent a family of transporters that are involved in the transport of glucose, amino acids, vitamins, and ions and other osmolytes across the brush-border membranes of kidney tubule cells and intestinal epithelial cells.

There are two SGLTs in the kidney involved in glucose reabsorption. SGLT1 is found primarily in the distal S2/S3 segment of the proximal tubule and SGLT2 is expressed exclusively in the S1 segment. The location of SGLT2 in the proximal tubule means that it is primarily responsible for glucose reabsorption. SGLT2 is a high-capacity low-affinity transporter that due to its expression location is responsible for approximately 90% of the glucose reabsorption activity of the kidney.

As would be expected from the name of the renal glucose transporters, SGLT1 and SGLT2 catalyse the active transport of glucose against a concentration gradient across the lumenal (apical) membrane of the tubule cell and couple this transport to sodium uptake. The inward sodium uptake is maintained by ATP-driven active transport of the sodium across the basolateral (anti-lumenal) membrane into the blood (coupled to inward uptake of potassium).

The reabsorbed glucose passively diffuses out of the tubule cell into the blood via the basolateral membrane associated GLUT2. Under normal conditions saturation of the ability of SGLT2 (and SGLT1) to reabsorb glucose is never saturated. The kidney can filter and reabsorb approximately 375mg of glucose per minute. The plasma concentration of glucose required to exceed this capacity is well above that considered normal and is only observed situations of renal dysfunction/disease or most importantly in type 2 diabetes.

Because of the importance of SGLT2 in renal reabsorption of glucose this transporter has become the target for therapeutic intervention of the hyperglycemia associated with type 2 diabetes. By specifically inhibiting SGLT2 there will be increased glucose excretion in the urine and thus a lowering of plasma glucose levels. Several SGLT2-specific inhibitors are currently in clinical trials with a few reaching phase III status.

Glucose Transporters

One major response of non-hepatic tissues to insulin is the recruitment, to the cell surface, of glucose transporter complexes. Glucose transporters comprise a family of at least 14 members. The most well characterized members of the family are GLUT1, GLUT2, GLUT3, GLUT4 and GLUT5. The glucose transporters are facilitative transporters that carry hexose sugars across the membrane without requiring energy. These transporters belong to a family of proteins called the solute carriers. Specifically, the official gene names for the GLUTs are solute carrier family 2 (facilitated glucose transporter) member. Thus, the GLUT1 gene symbol is SLC2A1, GLUT2 is SLC2A2, GLUT3 is SLC2A3, GLUT4 is SLC2A4 and GLUT5 is SLC2A5.

The glucose transporters can be divided into three classes based upon primary amino acid sequence comparisons. Class I transporters include GLUT1, GLUT2, GLUT3 (and the gene duplication of GLUT3 identified as GLUT14), and GLUT4. Class II transporters include GLUT5, GLUT7, GLUT9 and GLUT11. Class III transporters include GLUT6, GLUT8, GLUT10, GLUT12 and HMIT [proton (H^+) myoinositol symporter: SLC2A13]. HMIT is also known as GLUT13.

GLUT1 is ubiquitously distributed in various tissues with highest levels of expression seen in erythrocytes. In fact in erythrocytes GLUT1 accounts for almost 5% of total protein. Although widely expressed GLUT1 is not expressed in hepatocytes.

GLUT2 is found primarily in intestine, pancreatic β-cells, kidney and liver. The K_m of GLUT2 for glucose (17mM) is the highest of all the sugar transporters. The high K_m ensures a fast equilibrium of glucose between the cytosol and the extracellular space ensuring that liver and pancreas do not metabolize glucose until its levels rise sufficiently in the blood. GLUT2 molecules can transport both glucose and fructose. When the concentration of blood glucose increases in response to food intake, pancreatic GLUT2 molecules mediate an increase in glucose uptake which leads to increased insulin secretion. For this reason, GLUT2 is thought to be a "glucose sensor".

GLUT3 is found primarily in neurons but also found in the intestine. GLUT3 binds glucose with high affinity (has the lowest K_m of the GLUTs) which allows neurons to have enhanced access to glucose especially under conditions of low blood glucose.

Insulin-sensitive tissues, such as skeletal muscle and adipose tissue, contain GLUT4 whose mobilization to the cell-surface is stimulated by insulin action.

GLUT5 and the closely related transporter GLUT7 are involved in fructose transport. GLUT5 is expressed in intestine, kidney, testes, skeletal muscle, adipose tissue and brain. Although GLUT2, -5, -7, 8, -9, -11, and -12 can all transport fructose, GLUT5 is the only transporter that exclusively transports fructose.

GLUT9 (SLC2A9) does not transport sugar but is a uric acid transporter abundant in the kidney and liver.

Recent evidence has shown that the cell surface receptor for the human T cell leukemia virus (HTLV) is the ubiquitous GLUT1.

Fermentation

There are two important ways a cell can harvest energy from food:

fermentation and cellular respiration. Both start with the same first step: the process of glycolysis which is the breakdown or splitting of glucose (6 carbons) into two 3-carbon molecules called pyruvic acid. The energy from other sugars, such as fructose, is also harvested using this process. Glycolysis is probably the oldest known way of producing ATP. There is evidence that the process of glycolysis predates the existence of O_2 in the Earth's atmosphere and organelles in cells:

- Glycolysis does not need oxygen as part of any of its chemical reactions. It serves as a first step in a variety of both aerobic and anaerobic energy-harvesting reactions.
- Glycolysis happens in the cytoplasm of cells, not in some specialized organelle.

Glycolysis is the one metabolic pathway found in all living organisms.

Fermenation

Fermentation is the process of deriving energy from the oxidation of organic compounds, such as carbohydrates, and using an endogenous electron acceptor, which is usually an organic compound. In contrast, respiration is where electrons are donated to an exogenous electron acceptor, such as oxygen, via an electron transport chain.

Fermentation is important in anaerobic conditions when there is no oxidative phosphorylation to maintain the production of ATP (Adenosine triphosphate) by glycolysis. During fermentation, pyruvate is metabolised to various different compounds. Homolactic fermentation is the production of lactic acid from pyruvate; alcoholic fermentation is the conversion of pyruvate into ethanol and carbon dioxide; and heterolactic fermentation is the production of lactic acid as well as other acids and alcohols. Fermentation does not necessarily have to be carried out in an anaerobic environment. For example, even in the presence of abundant oxygen, yeast cells greatly prefer fermentation to oxidative phosphorylation, as long as sugars are readily available for consumption.

Sugars are the most common substrate of fermentation, and typical examples of fermentation products are ethanol, lactic acid, and hydrogen. However, more exotic compounds can be produced by fermentation, such as butyric acid and acetone. Yeast carries out fermentation in the production of ethanol in beers, wines and other alcoholic drinks, along with the production of large quantities of carbon dioxide. Fermentation occurs in mammalian muscle during periods of intense exercise where oxygen supply becomes limited, resulting in the creation of lactic acid.

Fermentation products contain chemical energy (they are not fully oxidized), but are considered waste products, since they cannot be metabolized further without the use of oxygen (or other more highly-oxidized electron acceptors). A consequence is that the production of adenosine triphosphate (ATP) by fermentation is less efficient than oxidative phosphorylation, whereby pyruvate is fully oxidized to carbon dioxide.

Ethanol Fermentation

Ethanol fermentation (performed by yeast and some types of bacteria) breaks the pyruvate down into ethanol and carbon dioxide. It is important in bread-making, brewing, and wine-making. Usually only one of the products is desired; in bread-making, the alcohol is baked out, and, in alcohol production, the carbon dioxide is released into the atmosphere or used for carbonating the beverage. When the ferment has a high concentration of pectin, minute quantities of methanol can be produced. The chemical equation below summarizes the fermentation of glucose, whose chemical formula is $C_6H_{12}O_6$. One glucose molecule is converted into two ethanol molecules and two carbon dioxide molecules:

$$C_6H_{12}O_6 \rightarrow 2C_2H_5OH + 2CO_2$$

C_2H_5OH is the chemical formula for ethanol.

Before fermentation takes place, one glucose molecule is broken down into two pyruvate molecules. This is known as glycolysis.

Lactic Acid Fermentation

Lactic acid fermentation is the simplest type of fermentation. Essentially, it is a redox reaction. In anaerobic conditions, the cell's primary mechanism of ATP production is glycolysis. Glycolysis reduces – transfers electrons to – NAD^+, forming NADH. However, there is only a limited supply of NAD^+ available in a cell. For glycolysis to continue, NADH must be oxidized – have electrons taken away – to regenerate the NAD^+. This is usually done through an electron transport chain in a process called oxidative phosphorylation; however, this mechanism is not available without oxygen.

Instead, the NADH donates its extra electrons to the pyruvate molecules formed during glycolysis. Since the NADH has lost electrons, NAD^+ regenerates and is again available for glycolysis. Lactic acid, for which this process is named, is formed by the reduction of pyruvate.

In heterolactic acid fermentation, one molecule of pyruvate is converted to lactate; the other is converted to ethanol and carbon dioxide. In homolactic acid fermentation, both molecules of pyruvate are converted to lactate. Homolactic acid fermentation is unique because it is one of the only respiration processes to not produce a gas as a by-product.

Homolactic fermentation breaks down the pyruvate into lactate. It occurs in the muscles of animals when they need energy faster than the blood can supply oxygen. It also occurs in some kinds of bacteria (such as lactobacilli) and some fungi. It is this type of bacteria that converts lactose into lactic acid in yogurt, giving it its sour taste.

These lactic acid bacteria can be classed as homofermentative, where the end product is mostly lactate, or heterofermentative, where some lactate is further metabolized and results in carbon dioxide, acetate or other metabolic products.

The process of lactic acid fermentation using glucose is summarized below. In homolactic fermentation, one molecule of glucose is converted to two molecules of lactic acid:

$$C_6H_{12}O_6 \rightarrow 2\ CH_3CHOHCOOH.$$

In heterolactic fermentation, the reaction proceeds as follows, with one molecule of glucose being converted to one molecule of lactic acid, one molecule of ethanol, and one molecule of carbon dioxide:

$$C_6H_{12}O_6 \rightarrow CH_3CHOHCOOH + C_2H_5OH + CO_2$$

Before lactic acid fermentation can occur, the molecule of glucose must be split into two molecules of pyruvate. This process is called glycolysis.

Glycolysis

To extract chemical energy from glucose, the glucose molecule must be split into two molecules of pyruvate. This process also generates two molecules of adenosine triphosphate as an immediate energy yield and two molecules of NADH.

$$C_6H_{12}O_6 + 2\ ADP + 2\ P_i + 2\ NAD^+ \rightarrow$$

$$2\ CH_3COCOO^- + 2\ ATP + 2\ NADH + 2\ H_2O + 2H^+$$

The chemical formula of pyruvate is CH_3COCOO^-. P_i stands for the inorganic phosphate. As shown by the reaction equation, glycolysis causes the reduction of two molecules of NAD^+ to NADH. Two ADF molecules are also converted to two ATP and two water molecules via substrate-level phosphorylation.

Aerobic Respiration

In aerobic respiration, the pyruvate produced by glycolysis is further oxidized completely, generating additional ATP and NADH in the citric acid cycle and by oxidative phosphorylation. However, this can only occur in the presence of oxygen. Oxygen is toxic to organisms which are obligate anaerobes, and are not required by facultative anaerobic organisms. In the absence of oxygen, one of the fermentation pathways occurs in order to regenerate NAD^+; lactic acid fermentation is one of these pathways.

Hydrogen Gas Production in Fermentation

Hydrogen gas is produced in many types of fermentation (mixed acid fermentation, butyric acid fermentation, caproate fermentation, butanol fermentation, glyoxylate fermentation), as a way to regenerate NAD^+ from NADH. Electrons are transferred to ferredoxin, which in turn is oxidized by hydrogenase, producing H_2.

Hydrogen gas is a substrate for methanogens and sulfate reducers, which keep the concentration of hydrogen sufficiently low to allow the production of such an energy-rich compound. In fermentation these pyruvic acid molecules are turned into some "waste" product, and a little bit of energy (only two ATP molecules per molecule of glucose – actually four are produced in glycolysis, but two are used up) is produced. Out of many possible types of fermentation processes, two of the most common types are lactic acid fermentation and

Lactic acid fermentation is done by some fungi, some bacteria like the *Lactobacillus acidophilus.* in yogurt, and sometimes by our muscles. Normally our muscles do cellular respiration like the rest of our bodies, using O_2 supplied by our lungs and blood. However, under greater exertion when the oxygen supplied by the lungs and blood system can't get there fast enough to keep up with the muscles' needs, our muscles can switch over and do lactic acid fermentation.

In the process of lactic acid fermentation, the 3-carbon pyruvic acid molecules are turned into lactic acid. It is the presence of lactic acid in yogurt that gives it its sour taste, and it is the presence of lactic acid in our muscles "the morning after" that makes them so sore. Once our muscles form lactic acid, they can't do anything else with it, so until it is gradually washed away by the blood stream and carried to the liver (which is able to get rid of it), our over-exerted muscles feel stiff and sore even if they haven't been physically injured.

Alcohol fermentation is done by yeast and some kinds of bacteria. The "waste" products of this process are ethanol and carbon dioxide (CO_2). Humans have long taken advantage of this process in making bread, beer, and wine. In bread making, it is the CO_2 which forms and is trapped between the gluten (a long protein in wheat) molecules that causes the bread to rise, and the ethanol (often abbreviated as EtOH – do you remember how to draw it?) evaporating that gives it its wonderful smell while baking. The effects of the ethanol in beer and wine are something with which many college students are familiar (sometimes too familiar?), and it is the CO_2 produced by the process of fermentation that makes these beverages effervescent.

Cellular Respiration

An analogy can be drawn between the process of cellular respiration in our cells and a car. The mitochondria are the engines of our cells where sugar is burned for fuel and the exhaust is CO_2 and H_2O. Note that in a car that burned fuel perfectly, the only exhaust should theoretically be CO_2 and H_2O also. There are three steps in the process of cellular respiration: glycolysis, the Krebs cycle, and the electron transport chain.

In contrast to fermentation, in the process of cellular respiration, the pyruvic acid molecules are broken down completely to CO_2 and more energy released. Note that three molecules of O_2 must react with each molecule of pyruvic acid to form the three carbon dioxide molecules, and three molecules of water are also formed to "use up" the hydrogens. As mentioned above, in glycolysis, a total of four molecules of ATP are produced, but two are used up in other steps in the process. Additional ATP is produced during the Krebs Cycle and the Electron Transport Chain, resulting in a grand total of 40 ATP molecules produced from the breakdown of one molecule of glucose via cellular respiration. Since two of those are used up during glycolysis, in prokaryotes a *net* total of 38 molecules of ATP are produced by cellular respiration.

Most prokaryotes have very simple cells which lack several types of organelles present in eukaryotes, and therefore the Krebs Cycle and the Electron Transport Chain occur in the cytoplasm and/or using chemicals embedded in the cell membrane. In contrast, eukaryotes have more complex cells with more specialized organelles to perform given functions. In eukaryotes, the Krebs Cycle and Electron Transport Chain occur within the mitochondria, and thus the pyruvic acid resulting from glycolysis must be sent into the mitochondria for these reactions to occur.

However, to move one molecule of pyruvic acid (remember each molecule of glucose turns into two pyruvic acid molecules) from the cytoplasm into a mitochondrion "costs" the cell one molecule of ATP (therefore two ATPs for a whole glucose), thus a *net* total of 36 ATP molecules per molecule of glucose is produced in eukaryotes as compared to only two in fermentation. The overall reaction for cellular respiration is $C_6H_{12}O_6 + 6O_2 \rightarrow 6CO_2 + 6H_2O$ (+ energy for the cell to use for other things).

In glycolysis and the Krebs cycle, there are also a number of electrons released as the glucose molecule is broken down. The cell must deal with these electrons in some way, so they are stored by the cell by forming a compound called NADH by the chemical reaction, $NAD^+ + H^+ + 2e^-$ NADH. This NADH is used to carry the electrons to the electron transport chain, where more energy is harvested from them.

In eukaryotes, the pyruvic acid from glycolysis must be transferred into the mitochondria to be sent through the Krebs cycle, also known as the citric acid cycle, at a "cost" of one ATP per molecule of pyruvic acid. In this cycle, discovered by Hans Krebs, the pyruvic acid molecules are converted to CO_2, and two more ATP molecules are produced per molecule of glucose. First, each 3-carbon pyruvic acid molecule has a CO_2 broken off and the other two carbons are transferred to a molecule called acetyl coenzyme A, while a molecule of NADH is formed from NAD^+ for each pyruvic acid (= 2 for the whole glucose). These acetyl CoA molecules are put into the actual cycle, and after the coenzyme A part is released, eventually each 2-carbon piece is broken apart into two molecules of CO_2. In the process, for each acetyl CoA that goes into the cycle, three molecules of NADH, one molecule of $FADH_2$, and one molecule of ATP are formed (= 6 NADH, 2 $FADH_2$, and 2 ATP per whole glucose).

The electron transport chain is a system of electron carriers embedded into the inner membrane of a mitochondrion. As electrons are passed from one compound to the next in the chain, their energy is harvested and stored by forming ATP. For each molecule of NADH which puts its two electrons in, approximately three molecules of ATP are formed, and for each molecule of $FADH_2$, about two molecules of ATP are formed.

Many of the compounds that make up the electron transport chain belong to a special group of chemicals called cytochromes. The central structure of a cytochrome is a porphyrin ring like chlorophyll but with iron in the centre (chlorophyll has magnesium). A porphyrin with iron in the centre is called a heme group, and these are also found in hemoglobin in our blood.

At the last step in the electron transport chain, the "used up" electrons, along with some "spare" hydrogen ions are combined with O_2 (we finally got around to the O_2) to form water as a waste product: $4e^- + 4H^+ + O_2$ $2H_2O$.

Many of the enzymes in the cells of organisms need other helpers to function. These non-protein enzyme helpers are called cofactors and can include substances like iron, zinc, or copper. If a cofactor is an organic molecule, it then is called a coenzyme. Many of the vitamins needed by our bodies are used as coenzymes to help our enzymes to do their jobs. Vitamin B_1 (thiamine) is a coenzyme used in removing CO_2 from various organic compounds. B_2 (riboflavin) is a component of FAD (or $FADH_2$), one of the chemicals used to transport electrons from the Krebs cycle to the electron transport chain.

Vitamin B_3 (niacin) is a component of NAD^+ (or NADH) which is the major transporter of electrons from glycolysis and the Krebs cycle to the electron transport chain. Without enough of these B vitamins, our ability to get the energy out of our food would come to a grinding halt! B_6 (pyridoxine), B_{12} (cobalamin), pantothenic acid, folic acid, and biotin are all other B vitamins which serve as coenzymes at various points in metabolizing our food. Interestingly, B_{12} has cobalt in it, a mineral which we need in only very minute quantities, but whose absence can cause symptoms of deficiency.

Gluconeogenesis

Gluconeogenesis is the biosynthesis of new glucose, (i.e. not glucose from glycogen). The production of glucose from other metabolites is necessary for use as a fuel source by the brain, testes, erythrocytes and kidney medulla since glucose is the sole energy source for these organs. During starvation, however, the brain can derive energy from ketone bodies which are converted to acetyl-CoA. The primary carbon skeletons used for gluconeogenesis are derived from pyruvate, lactate, glycerol, and the amino acids alanine and glutamine.

The relevant reactions of gluconeogenesis are depicted. The enzymes of the 3 bypass steps are indicated in green along with phosphoglycerate kinase. This latter enzyme is included since when functioning in the gluconeogenic direction the reaction consumes energy. Gluconeogenesis from 2 moles of pyruvate to 2 moles of 1,3-bisphosphoglycerate consumes 6 moles of ATP. This makes the process of gluconeogenesis very costly from an energy standpoint considering that glycolysis to 2 moles of pyruvate only yields 2 moles of ATP. Note that several steps are required in going from 2 moles of 1,3-bisphosphoglycerate to 1 mole of fructose-1,6-bisphosphate.

First there is a reversal of the glyceraldehyde-3-phosphate dehydrogenase reaction which requires a supply of NADH. When lactate is the gluconeogenic substrate the NADH is supplied by the lactate dehydrogenase reaction, and it is supplied by the malate dehydrogenase reaction when pyruvate is the substrate. Secondly, 1 mole of glyceraldehyde-3-phosphate must be isomerized to DHAP and then a mole of DHAP can be condensed to a mole of glyceraldehyde-3-phosphate to form 1 mole of fructose-1,6-bisphosphate in a reversal of the aldolase reaction.

Most non-hepatic tissues lack glucose-6-phosphatase and so the glucose-6-phosphate generated in these tissues would be a substrate for glycogen synthesis. In hepatocytes the glucose-6-phosphatase reactions allows the liver to supply the blood with free glucose. Remember that due to the high K_m of liver glucokinase most of the glucose will not be phosphorylated and will flow down its' concentration gradient out of hepatocytes and into the blood.

The three reactions of glycolysis that proceed with a large negative free energy change are bypassed during gluconeogenesis by using different enzymes. These three are the pyruvate kinase, phosphofructokinase-1(PFK-1) and hexokinase/glucokinase catalysed reactions. In the liver or kidney cortex and in some cases skeletal muscle, the glucose-6-phosphate (G6P) produced by gluconeogenesis can be incorporated into glycogen. In this case the third bypass occurs at the glycogen phosphorylase catalysed reaction. Since skeletal muscle lacks glucose-6-phosphatase it cannot deliver free glucose to the blood and undergoes gluconeogenesis exclusively as a mechanism to generate glucose for storage as glycogen.

Pyruvate to Phosphoenolpyruvate (PEP), Bypass 1

Conversion of pyruvate to PEP requires the action of two mitochondrial enzymes. The first is an ATP-requiring reaction catalysed by pyruvate carboxylase, (PC). As the name of the enzyme implies, pyruvate is carboxylated to form oxaloacetate (OAA). The CO_2 in this reaction is in the form of bicarbonate (HCO_3^-). This reaction is an anaplerotic reaction since it can be used to fill-up the TCA cycle.

The second enzyme in the conversion of pyruvate to PEP is PEP carboxykinase (PEPCK). PEPCK requires GTP in the decarboxylation of OAA to yield PEP. Since PC incorporated CO_2 into pyruvate and it is subsequently released in the PEPCK reaction, no net fixation of carbon occurs. Human cells contain almost equal amounts of mitochondrial and cytosolic PEPCK so this second reaction can occur in either cellular compartment.

For gluconeogenesis to proceed, the OAA produced by PC needs to be transported to the cytosol. However, no transport mechanism exist for its' direct transfer and OAA will not freely diffuse. Mitochondrial OAA can become cytosolic via three pathways, conversion to PEP (as indicated above through the action of the mitochondrial PEPCK), transamination to aspartate or reduction to malate, all of which are transported to the cytosol.

If OAA is converted to PEP by mitochondrial PEPCK, it is transported to the cytosol where it is a direct substrate for gluconeogenesis and nothing further is required. Transamination of OAA to aspartate allows the aspartate to be transported to the cytosol where the reverse transamination occurs yielding cytosolic OAA. This transamination reaction requires continuous transport of glutamate into, and α-ketoglutarate out of, the mitochondrion. Therefore, this process is limited by the availability of these other substrates. Either of these latter two reactions will predominate when the substrate for gluconeogenesis is lactate. Whether mitochondrial decarboxylation or transamination occurs is a function of the availability of PEPCK or transamination intermediates.

Mitochondrial OAA can also be reduced to malate in a reversal of the TCA cycle reaction catalysed by malate dehydrogenase (MDH). The reduction of OAA to malate requires NADH, which will be accumulating in the mitochondrion as the energy charge increases. The increased energy charge will allow cells to carry out the ATP costly process of gluconeogenesis. The resultant malate is transported to the cytosol where it is oxidized to OAA by cytosolic MDH which requires NAD^+ and yields NADH. The NADH produced during the cytosolic oxidation of malate to OAA is utilized during the glyceraldehyde-3-phosphate dehydrogenase reaction of glycolysis. The coupling of these two oxidation-reduction reactions is required to keep gluconeogenesis functional when pyruvate is the principal source of carbon atoms. The conversion of OAA to malate predominates when pyruvate (derived from glycolysis or amino acid catabolism) is the source of carbon atoms for gluconeogenesis. When in the cytoplasm, OAA is converted to PEP by the cytosolic version of PEPCK. Hormonal signals control the level of PEPCK protein as a means to regulate the flux through gluconeogenesis.

The net result of the PC and PEPCK reactions is:

Pyruvate + ATP + GTP + H_2O $\rightarrow$

PEP + ADP + GDP + P_i + $2H^+$

Fructose-1,6-bisphosphate to Fructose-6-phosphate, Bypass 2

Fructose-1,6-bisphosphate (F1,6BP) conversion to fructose-6-phosphate (F6P) is the reverse of the rate limiting step of glycolysis. The reaction, a simple hydrolysis, is catalysed by fructose-1,6-bisphosphatase (F1,6BPase). Like the regulation of glycolysis occurring at the PFK-1 reaction, the F1,6BPase reaction is a major point of control of gluconeogenesis.

Glucose-6-phosphate (G6P) to Glucose (or Glycogen), Bypass 3

G6P is converted to glucose through the action of glucose-6-phosphatase (G6Pase). This reaction is also a simple hydrolysis reaction like that of F1,6BPase. Since the brain and skeletal muscle, as well as most non-hepatic tissues, lack G6Pase activity, any gluconeogenesis that occurs in these tissues is not utilized for blood glucose supply. In the kidney, muscle and especially the liver, G6P be shunted toward glycogen if blood glucose levels are adequate. The reactions necessary for glycogen synthesis are an alternate bypass series of reactions. Phosphorolysis of glycogen is carried out by glycogen phosphorylase, whereas, glycogen synthesis is catalysed by glycogen synthase. The G6P produced from gluconeogenesis can be converted to glucose-1-phosphate (G1P) by phosphoglucose mutase (PGM). G1P is then converted to UDP-glucose (the substrate for glycogen synthase) by UDP-glucose pyrophosphorylase, a reaction requiring hydrolysis of UTP.

Substrates for Gluconeogenesis

Lactate: Lactate is a predominate source of carbon atoms for glucose synthesis by gluconeogenesis. During anaerobic glycolysis in skeletal muscle, pyruvate is reduced to lactate by lactate dehydrogenase (LDH). This reaction serves two critical functions during anaerobic glycolysis. First, in the direction of lactate formation the LDH reaction requires NADH and yields NAD^+ which is then available for use by the glyceraldehyde-3-phosphate dehydrogenase reaction of glycolysis. These two reaction are, therefore, intimately coupled during anaerobic glycolysis. Secondly, the lactate produced by the LDH reaction is released to the blood stream and transported to the liver where it is converted to glucose. The glucose is then returned to the blood for use by muscle as an energy source and to replenish glycogen stores. This cycle is termed the Cori cycle.

Pyruvate

Pyruvate, generated in muscle and other peripheral tissues, can be transaminated to alanine which is returned to the liver for gluconeogenesis. The transamination reaction requires an α-amino acid as donor of the amino group, generating an α-keto acid in the process. This pathway is termed the glucose-alanine cycle. Although the majority of amino acids are degraded in the liver some are deaminated in muscle.

The glucose-alanine cycle is, therefore, an indirect mechanism for muscle to eliminate nitrogen while replenishing its energy supply. However, the major function of the glucose-alanine cycle is to allow non-hepatic tissues to deliver the amino portion of catabolized amino acids to the liver for excretion as urea. Within the liver the alanine is converted back to pyruvate and used as a gluconeogenic substrate (if that is the hepatic requirement) or oxidized in the TCA cycle. The amino nitrogen is converted to urea in the urea cycle and excreted by the kidneys.

The glucose-alanine cycle is used primarily as a mechanism for skeletal muscle to eliminate nitrogen while replenishing its energy supply. Glucose oxidation produces pyruvate which can undergo transamination to alanine. This reaction is catalysed by alanine transaminase, ALT (ALT used to be referred to a serum glutamate-pyruvate transaminase, SGPT). Additionally, during periods of fasting, skeletal muscle protein is degraded for the energy value of the amino acid carbons and alanine is a major amino acid in protein. The alanine then enters the blood stream and is transported to the liver. Within the liver alanine is converted back to pyruvate which is then a source of carbon atoms for gluconeogenesis. The newly formed glucose can then enter the blood for delivery back to the muscle. The amino group transported from the muscle to the liver in the form of alanine is converted to urea in the urea cycle and excreted.

Amino Acids

All 20 of the amino acids, excepting leucine and lysine, can be degraded to TCA cycle intermediates as discussed in the metabolism of amino acids. This allows the carbon skeletons of the amino acids to be converted to those in oxaloacetate and subsequently into pyruvate. The pyruvate thus formed can be utilized by the gluconeogenic pathway. When glycogen stores are depleted, in muscle during exertion and liver during fasting, catabolism of muscle proteins to amino acids contributes the major source of carbon for maintenance of blood glucose levels.

Glycerol

Oxidation of fatty acids yields enormous amounts of energy on a molar basis, however, the carbons of the fatty acids cannot be utilized for net synthesis of glucose. The two carbon unit of acetyl-CoA derived from β-oxidation of fatty acids can be incorporated into the TCA cycle, however, during the TCA cycle two carbons are lost as CO_2. Thus, explaining why fatty acids do not undergo net conversion to carbohydrate.

The glycerol backbone of lipids can be used for gluconeogenesis. This requires phosphorylation to glycerol-3-phosphate by glycerol kinase and dehydrogenation to dihydroxyacetone phosphate (DHAP) by glyceraldehyde-3-phosphate dehydrogenase (G3PDH). The G3PDH reaction is the same as that used in the transport of cytosolic reducing equivalents into the mitochondrion for use in oxidative phosphorylation. This transport pathway is called the glycerol-phosphate shuttle.

The glycerol phosphate shuttle is a secondary mechanism for the transport of electrons from cytosolic NADH to mitochondrial carriers of the oxidative phosphorylation pathway. The primary cytoplasmic NADH electron shuttle is the malate-aspartate shuttle. Two enzymes are involved in this shuttle. One is the cytosolic version of the enzyme glycerol-3-phosphate dehydrogenase (glycerol-3-PDH) which has as one substrate, NADH. The second is the mitochondrial form of the enzyme which has as one of its' substrates, FAD^+.

The net result is that there is a continual conversion of the glycolytic intermediate, DHAP and glycerol-3-phosphate with the concomitant transfer of the electrons from reduced cytosolic NADH to mitochondrial oxidized FAD^+. Since the electrons from mitochondrial $FADH_2$ feed into the oxidative phosphorylation pathway at coenzyme Q (as opposed to NADH-ubiquinone oxidoreductase [complex I]) only 2 moles of ATP will be generated from glycolysis. G3PDH is glyceraldehyde-3-phoshate dehydrogenase.

The glycerol backbone of adipose tissue stored triacylgycerols is ensured of being used as a gluconeogenic substrate since adipose cells lack glycerol kinase. In fact adipocytes require a basal level of glycolysis in order to provide them with DHAP as an intermediate in the synthesis of triacyglycerols.

Propionate

Oxidation of fatty acids with an odd number of carbon atoms and the oxidation of some amino acids generates as the terminal oxidation product, propionyl-CoA. Propionyl-CoA is converted to the TCA intermediate, succinyl-CoA. This conversion is carried out by the ATP-requiring enzyme, propionyl-CoA carboxylase then methylmalonyl-CoA epimerase and finally the vitamin B_{12} requiring enzyme, methylmalonyl-CoA mutase. The utilization of propionate in gluconeogenesis only has quantitative significance in ruminants.

Role of Renal Gluconeogenesis

Although the liver has the critical role of maintaining blood glucose homeostasis and therefore, is the major site of gluconeogenesis, the kidney plays an important role. During periods of severe hypoglycemia that occur under conditions of hepatic failure, the kidney can provide glucose to the blood via renal gluconeogenesis. In the renal cortex, glutamine is the preferred substance for gluconeogenesis.

Glutamine is produced in high amounts by skeletal muscle during periods of fasting as a means to export the waste nitrogen resulting from amino acid catabolism. Through the actions of transaminases, a mole of waste ammonia is transferred to α-ketoglutarate via the glutamate dehydrogenase catalysed reaction yielding glutamate. Glutamate is then a substrate for glutamine synthetase which incorporates another mole of waste ammonia generating glutamine.

The glutamine is then transported to the kidneys where the reverse reactions occur liberating the ammonia and producing α-ketoglutarate which can enter the TCA cycle and the carbon atoms diverted to gluconeogenesis via oxaloacetate. This process serves two important functions. The ammonia (NH_3) that is liberated spontaneously ionizes to ammonium ion (NH_4^+) and is excreted in the urine effectively buffering the acids in the urine. In addition, the glucose that is produced via gluconeogenesis can provide the brain with critically needed energy.

Regulation of Gluconeogenesis

Obviously the regulation of gluconeogenesis will be in direct contrast to the regulation of glycolysis. In general, negative effectors of glycolysis are positive effectors of gluconeogenesis. Regulation of the activity of PFK-1 and F1,6BPase is the most significant site for controlling the flux toward glucose oxidation or glucose synthesis. As described in control of glycolysis, this is predominantly controlled by fructose-2,6-bisphosphate, F2,6BP which is a powerful negative allosteric effector of F1,6Bpase activity.

Regulation of glycolysis and gluconeogenesis by fructose 2,6-bisphosphate (F2,6BP). The major sites for regulation of glycolysis and gluconeogenesis are the phosphofructokinase-1 (PFK-1) and fructose-1,6-bisphosphatase (F-1,6-BPase) catalysed reactions. PFK-2 is the kinase activity and F-2,6-BPase is the phosphatase activity of the bi-functional regulatory enzyme, phosphofructokinase-2/fructose-2,6-bisphosphatase. PKA is cAMP-dependent protein kinase which phosphorylates PFK-2/F-2,6-BPase turning on the phosphatase activity. (+ve) and (-ve) refer to positive and negative activities, respectively.

The level of F2,6BP will decline in hepatocytes in response to glucagon stimulation as well as stimulation by catecholamines. Each of these signals is elicited through activation of cAMP-dependent protein kinase (PKA). One substrate for PKA is PFK-2, the bifunctional enzyme responsible for the synthesis and hydrolysis of F2,6BP. When PFK-2 is phosphorylated by PKA it acts as a phosphatase leading to the dephosphorylation of F2,6BP with a concomitant increase in F1,6Bpase activity and a decrease in PFK-1 activity. Secondarily, F1,6Bpase activity is regulated by the ATP/ADP ratio. When this is high, gluconeogenesis can proceed maximally.

Gluconeogenesis is also controlled at the level of the pyruvate to PEP bypass. The hepatic signals elicited by glucagon or epinephrine lead to phosphorylation and inactivation of pyruvate kinase (PK) which will allow for an increase in the flux through gluconeogenesis.

PK is also allosterically inhibited by ATP and alanine. The former signals adequate energy and the latter that sufficient substrates for gluconeogenesis are available. Conversely, a reduction in energy levels as evidenced by increasing concentrations of ADP lead to inhibition of both PC and PEPCK. Allosteric activation of PC occurs through acetyl-CoA. Each of these regulations occurs on a short time scale, whereas long-term regulation can be effected at the level of PEPCK. The amount of this enzyme increases in response to prolonged glucagon stimulation. This situation would occur in a starving individual or someone with an inadequate diet.

Whereas glucagon actions results in increased levels of cAMP and subsequent activation of gluconeogenesis, insulin action exerts the opposite effect. The mechanisms by which insulin turns off gluconeogenesis are complex. Reduction in the level of cAMP is exerted via the insulin-mediated activation of phosphodiesterase which hydrolyses cAMP to AMP. At the level of the regulation of genes involved in gluconeogenesis, cAMP signalling leads to phosphorylation of the transcription factor CREB at Ser133.

When phospho-CREB binds to the cAMP response element (CRE) of a target gene it results in the recruitment of the coactivators CBP and p300 (which are closely related). This complex activates gene expression through their intrinsic histone acetyltransferase activity and through recruitment of other coactivator molecules. The coactivator CBP is a target of insulin-dependent phosphorylation at Ser436. This residue, which is adjacent to the CREB-binding domain (CREB-BD), is phosphorylated by a phosphoinositol-3-kinase (PI3K)"dependent insulin signalling pathway, and is not conserved in the related cofactor p300. The protein coactivator PGC-1α (peroxisome proliferator-activated receptor-γ coactivator-1α) functions as a central regulator of gluconeogenesis by binding to several factors, including the glucocorticoid receptor, hepatic nuclear factor 4α (HNF4α) and a forkhead transcription factor (FOXO1) upstream of several genes encoding gluconeogenic enzymes. One mechanism by which insulin signalling antagonizes gluconeogenesis is through phosphorylation of FOXO1 and its subsequent exclusion from the nucleus. Expression of PGC-1α, however, is also regulated by cAMP through a well-defined CRE present in the promoter region.

Bibliography

Alka Pareek: *Environment and Nutritional Disorders*, Aavishkar Pub, Delhi, 2003.

Antwerpen, F.J. : *The Origins of Chemical Engineering, History of Chemical Engineering*, Washington D.C.: ACS, 1980.

Anupama Rani: *Assessment of Nutritional Status*, Sonali Publications, Delhi, 2010.

Arvind N. : *Inorganic Biochemistry,* Discovery, Delhi, 2010.

Ashraf, Syed Mahboob: *Handbook of Fruit and Vegetable Products,* Agrobios, Delhi, 2008.

Bagdi, Madan Lal : *Physiology, Biochemistry and Biotechnology,* Manglam Pub, Delhi, 2007.

Banerjee, S : *Biochemistry and Biotechnology,* Dominant Pub, Delhi, 2009.

Barrington, E. J. W.: *Biochemistry of Primitive Deuterostomians*, London, Academic Press, 1974.

Bhaskar, Harsh : *Basic Facts on Biochemistry*, Campus Books, Delhi, 2010.

Bhatia, A L : *Biochemistry and Endocrinology,* Indus Valley, Delhi, 2002.

Bhatnagar, Vasudev : *Cell Science and Technology*, Campus Books, Delhi, 2009.

Bora, K.K. : *Agros Dictionary of Plant Physiology and Biochemistry*, Agrobios, Delhi, 2001.

Bose, N. Sarath Chandra : *Biochemistry : A Practical Manual,* Pharma Med Press, Delhi, 2010.

Bose, Sarath Chandra : *Biochemistry : A Practical Manual*, Pharma Med Press, Delhi, 2010.

Cao Van P.: *An Integrated Approach for the Production and Processing of Minor Fruits in Martinique (FWI)*, Port of Spain, Trinidad and Tobago, 1996.

Chakraborty, Chiranjib : *Advances in Biochemistry and Biotechnology,* Daya, Delhi, 2005.

Chambers, K. L.: *Biochemical Coevolution: Eugene,* Oregon State University Press, 1970.

Chanderlekha Goswami: *Biochemistry and Instrumentation,* Manglam Pub, Delhi, 2011.

Chauhan, B.S. : *Principles of Biochemistry and Biophysics,* Laxmi Publications, Delhi, 2008.

Chauhan, Sushil : *Experiments in Biochemistry,* Campus Books, Delhi, 2010.

Chiranjib Chakraborty: *Advances in Biochemistry and Biotechnology,* Daya, Delhi, 2005.

Collymore L.: *Fruit Production in Barbados,* Port of Spain, Trinidad and Tobago, 1996.

Coultate, T. P. : *Food: The Chemistry of its Components,* London: The Royal Society of Chemistry, 1988.

Daphne C. Elliott: *Biochemistry and Molecular Biology,* Oxford University Press, Delhi, 2005.

Dar, Ghulam Hassan : *Soil Microbiology and Biochemistry,* New India Publishing Agency, Delhi, 2010.

Devi, C.R. Sudharmai : *Analytical Procedures in Soil Science and Agricultural Chemistry,* Agrotech, Delhi, 2004.

Dhatt A. S. and Singh Z.: *Propagation and Rootstocks of Citrus,* New Delhi, Malhotra Publishers, 1993.

Dhruve, J J : *Biochemistry at a Glance,* Agrotech Pub, Delhi, 2007.

Dutta, Chiranjib : *Fundamentals of Biochemistry : A Practical Approach,* Kanishka, Delhi, 2005.

Dutta, Naren Kumar : *Fundamentals of Biochemistry : A Practical Approach,* Kanishka, Delhi, 2005.

Elliott, B N : *Biochemistry and Molecular Biology,* Oxford University Press, Delhi, 2005.

Florkin, M.: *Comprehensive Biochemistry: Amsterdam,* Elsevier, 1975.

Frayn, K N : *Lipid Biochemistry,* Blackwell Science, London, 2004.

Ganguly, Smriti : *Biochemistry of Biomolecules,* Pearl Books, Delhi, 2007.

Garg, G.P. : *Biochemistry of Catalysts,* Discovery Pub, Delhi, 2010.

Goel, A K : *Basic Concept of Organic Chemistry,* Pearl Books, Delhi, 2008.

Golakiya, B A : *DNA : A Bridge Between Biochemistry and Biotechnology,* New India Publishing Agency, Delhi, 2008.

Goswami, Chanderlekha : *Biochemistry and Instrumentation,* Manglam Pub, Delhi, 2011.

Harriott, P. : *Unit Operations of Chemical Engineering,* 7th Edition, McGraw Hill, 2004.

Hemantaranjan, A L : *Developments in Physiology, Biochemistry and Molecular Biology of Plants,* New India Pub, Delhi, 2005.

Jasra, O.P. : *Environmental Biochemistry,* Sarup & Sons, Delhi, 2002.

Khalsa, Nisha : *Essentials of Biochemistry,* Aavishkar Pub, Delhi, 2008.

Khanna, V K : *Objective Genetics, Biotechnology, Biochemistry and Forestry,* I.K. International Publishing House, Delhi, 2008.

Kumar, Anil : *A Comprehensive Text Book of Applied Chemistry,* Abhishek Pub, Delhi, 2009.

Kumar, Vijay : *Insect Biochemistry,* Vista International Publishing House, Delhi, 2008.

Mahajan, Ritu : *Practical Biochemistry : Laboratory Mannual for Pharmacy Students,* Vayu Education of India, 2009.

Mahaling, Ram Naresh : *Basics of Biochemistry,* Anmol, Delhi, 2008.

Mark, Herman F. : *Polymer Chemistry: The Past 100 Years, Chemical & Engineering News,* April 6, 1976.

Matsuoka H.: *Cultivation of Panicum Genetic Resources for Evaluation of Characteristics,* Tokyo, JICA, 1997.

Meena Francis: *Biotech's Dictionary of Biochemistry,* Biotech Books, Delhi, 2007.

Mido, Y. : *An Introduction to Nuclear Chemistry,* Discovery, Delhi, 1995.

Narvekar, Raghunath : *Handbook of Biochemistry,* Adhyayan, Delhi, 2008.

Narvekar, Raghunath : *Molecular Biochemistry : Principles and Practices,* Adhyayan Pub, Delhi, 2008.

Nissen, J. : *Enzymic Hydrolysis of Food Proteins,* London: Elsevier Applied Science, 1985.

Ojha, Jai Shankar : *Aquaculture and Nutrition and Biochemistry,* Agrotech, Delhi, 2006.

Pandey, Ashok : *Biotechnology : Food Fermentation (Microbiology, Biochemistry and Technology)* , Educational Publishers, Delhi, 1999.

Pani, Balram : *Textbook of Environmental Chemistry,* I K International, Delhi, 2007.

Patricia Trueman: *Nutritional Biochemistry*, MJP Pub, Delhi, 2007.

Paul M. Althouse: *Introduction to Agricultural Biochemistry*, Biotech Books, Delhi, 2005.

Prakash, M : *Insect Biochemistry*, Discovery Pub, Delhi, 2008.

Prasad, G.P. : *Biochemistry of Proteins*, Discovery, Delhi, 2010.

Prasad, S.K. : *Biochemistry of Carbohydrates*, Discovery Pub, Delhi, 2010.

Price, C. C.: *Synthesis of Life: Stroudsburg*, Pa., Dowden, Hutchinson & Ross, 1974.

Purohit, S.S. : *Biochemistry : Fundamentals and Applications*, Agrobios, Delhi, 2009.

Rastogi, Arun : *An Introduction to Practical Biochemistry*, Anmol, Delhi, 2010.

Reilly, P. J.: *Enzymic degradation of starch, In Starch conversion technology*, New York: Marcel Dekker Inc, 1984.

Samuel Delvin: *Science and Technology : Biochemistry*, Sarup, Delhi, 2007.

Sangha, S P S : *Basic Techniques in Biochemistry and Molecular Biology*, I K International, Delhi, 2009.

Sarkar, Arun : *Advanced Organic Chemistry : Reactions and Mechanisms*, Swastik Publications, Delhi, 2011.

Satake, M. : *An Introduction to Analytical Chemistry*, Discovery, Delhi, 1999.

Sharma, Anil : *Cell Biology, Genetics and Plant Breeding*, Indus Valley Pub, Delhi, 2006.

Sharma, Pradeep : *Biochemistry and Organisation of Cells*, RBSA Pub, Delhi, 2006.

Sharma, R K : *Chemistry of Chemical Bonding*, Discovery, Delhi, 2007.

Sharma, Pradeep : *Biochemistry and Organisation of Cells*, BSA Pub, Delhi, 2006.

Shukla, Arvind N. : *Inorganic Biochemistry*, Discovery, Delhi, 2010.

Shukla, N. : *Inorganic Biochemistry*, Discovery, Delhi, 2010.

Siddhu, Anupa : *Qualitative Tests and Quantitative Procedures in Biochemistry*, Phoenix Pub, Delhi, 2002.

Singh, Bimla : *Biochemistry and Genomic Revolution*, Vista International Pub, Delhi, 2006.

Singh, G : *Chemistry of Amino-Acids and Proteins*, Discovery, Delhi, 2007.

Singh, Mahinder : *A Textbook of Biochemistry,* Dominant Pub, Delhi, 2011.

Sinha G. C., Reddy Y. T. N. and Singh G.: *Propagation and Rootstocks in Guava,* New Delhi, Malhotra Publishers, 1993.

Smriti Ganguly: *Biochemistry of Biomolecules,* Pearl Books, Delhi, 2007.

Srivastava, Seema : *Plant Physiology and Biochemistry,* Campus Books, Delhi, 2009.

Sunetra Roday: *Food Science and Nutrition,* Oxford University Press, Delhi, 2007.

Suri, Nitin : *Molecular Biology and Biochemistry,* Oxford Book Company, Delhi, 2010.

Sushil Chauhan: *Experiments in Environmental Chemistry,* Campus Books, Delhi, 2010.

Swarnim, K. : *A Textbook of Biochemistry and Microbiology,* Surendra Pub, Delhi, 2010.

Tiwari, B N : *Diet and Nutrition,* Pearl Books, Delhi, 2007.

Tripathi, Pramila: *Botanical Pesticides in the Management of Post-harvest Fruit Diseases,* Daya, Delhi, 2005.

Valli Manickam: *A Textbook of Analytical Chemistry,* PharmaMed Press, Delhi, 2006.

Whyman, Robin : *Applied Organometallic Chemistry and Catalysis,* Oxford University Press, Delhi, 2005.

Yadav, M : *Nutritional Biochemistry and Metabolism,* Arise Pub, Delhi, 2008.

Index

□□□